INTRODUCTION TO LINEAR ALGEBRA

INTRODUCTION TO LINEAR ALGEBRA

FOURTH EDITION

STUDENT'S SOLUTIONS MANUAL

JIMMY T. ARNOLD

LEE W. JOHNSON

R. DEAN RIESS

Virginia Polytechnic Institute and State University

An imprint of Addison Wesley Longman, Inc.

Reading, Massachusetts • Menlo Park, California • New York • Harlow, England
Don Mills, Ontario • Sydney • Mexico City • Madrid • Amsterdam

Reproduced by Addison-Wesley Publishing Company Inc. from camera-ready copy supplied by the author.

ISBN 0-201-44276-0

2 3 4 5 6 7 8 9 10 VG 0099

Preface

The Student Solutions Manual provides expanded solutions to the odd-numbered computational exercises given in the text. It includes all the exercises for which an answer is provided in the answer key in the back of the text. Thus, the manual supplements the answer key by showing the essential steps in the solution process. The manual also includes solutions to selected odd-numbered exercises, not included in the answer key. As a general rule, solutions that involve proofs are not included.

Contents

Chapter 1

Matrices and Systems of Equations

1.1 Introduction to Matrices and Systems of Linear Equations

1. Linear.

3. Linear.

5. Nonlinear.

7. $$\begin{aligned} x_1 + 3x_2 &= 7 & 1 + 3 \cdot 2 &= 7 \\ 4x_1 - x_2 &= 2 & 4 \cdot 1 - 2 &= 2 \end{aligned}$$

9. $$\begin{aligned} x_1 + x_2 &= 0 & 1 + (-1) &= 0 \\ 3x_1 + 4x_2 &= -1 & 3 \cdot 1 + 4 \cdot (-1) &= -1 \\ -x_1 + 2x_2 &= -3 & -1 + 2 \cdot (-1) &= -3 \end{aligned}$$

11. Unique solution.

13. Infinitely many solutions.

15. (a) The planes do not intersect; that is, the planes are parallel.

    (b) The planes intersect in a line or the planes are coincident.

17. The planes intersect in the line $x = 4 - 3t, y = 2t - 1, z = t$.

19. $A = \begin{bmatrix} 2 & 1 & 6 \\ 4 & 3 & 8 \end{bmatrix}$.

21. $Q = \begin{bmatrix} 1 & 4 & -3 \\ 2 & 1 & 1 \\ 3 & 2 & 1 \end{bmatrix}$.

23. $\begin{array}{rcrcl} 2x_1 & + & x_2 & = & 6 \\ 4x_1 & + & 3x_2 & = & 8 \end{array}$; $\begin{array}{rcrcr} x_1 & + & 4x_2 & = & -3 \\ 2x_1 & + & x_2 & = & 1 \\ 3x_1 & + & 2x_2 & = & 1 \end{array}$

25. $A = \begin{bmatrix} 1 & 1 & -1 \\ 2 & 0 & -1 \end{bmatrix}$, $B = \begin{bmatrix} 1 & 1 & -1 & 2 \\ 2 & 0 & -1 & 1 \end{bmatrix}$.

27. $A = \begin{bmatrix} 1 & 1 & 2 \\ 3 & 4 & -1 \\ -1 & 1 & 1 \end{bmatrix}$, $B = \begin{bmatrix} 1 & 1 & 2 & 6 \\ 3 & 4 & -1 & 5 \\ -1 & 1 & 1 & 2 \end{bmatrix}$.

29. $A = \begin{bmatrix} 1 & 1 & 1 \\ 2 & 3 & 1 \\ 1 & -1 & 3 \end{bmatrix}$, $B = \begin{bmatrix} 1 & 1 & 1 & 1 \\ 2 & 3 & 1 & 2 \\ 1 & -1 & 3 & 2 \end{bmatrix}$.

31. Elementary operations on equations: $E_2 - E_1,\ E_3 + 2E_1$.

Reduced system of equations: $\begin{array}{rcl} x_1 + 2x_2 - x_3 & = & 1 \\ -x_2 + 3x_3 & = & 1 \\ 5x_2 - 2x_3 & = & 6 \end{array}$.

Elementary row operations: $R_2 - R_1,\ R_3 + 2R_1$.

Reduced augmented matrix: $\begin{bmatrix} 1 & 2 & -1 & 1 \\ 0 & -1 & 3 & 1 \\ 0 & 5 & -2 & 6 \end{bmatrix}$.

33. Elementary operations on equations: $E_2 - E_1,\ E_3 - 3E_1$.

Reduced system of equations: $\begin{array}{rcr} x_1 + x_2 & = & 9 \\ -2x_2 & = & -2 \\ -2x_2 & = & -21 \end{array}$.

Elementary row operations: $R_2 - R_1,\ R_3 - 3R_1$.

Reduced augmented matrix: $\begin{bmatrix} 1 & 1 & 9 \\ 0 & -2 & -2 \\ 0 & -2 & -21 \end{bmatrix}$.

35. Elementary operations on equations: $E_2 \leftrightarrow E_1,\ E_3 + E_1$.

Reduced system of equations: $\begin{array}{rcl} x_1 + 2x_2 - x_3 + x_4 & = & 1 \\ x_2 + x_3 - x_4 & = & 3 \\ 3x_2 + 6x_3 & = & 1 \end{array}$.

Elementary row operations: $R_2 \leftrightarrow R_1,\ R_3 + R_1$.

Reduced augmented matrix: $\begin{bmatrix} 1 & 2 & -1 & 1 & 1 \\ 0 & 1 & 1 & -1 & 3 \\ 0 & 3 & 6 & 0 & 1 \end{bmatrix}$.

1.2 Echelon Form and Gauss-Jordan Elimination

1. The matrix is in echelon form. The row operation $R_2 - 2R_1$ transforms the matrix to reduced echelon form $\begin{bmatrix} 1 & 0 \\ 0 & 1 \end{bmatrix}$.

3. Not in echelon form. $(1/2)R_1$, $R_2 - 4R_1$, $(-1/5)R_2$ yields echelon form $\begin{bmatrix} 1 & 3/2 & 1/2 \\ 0 & 1 & 2/5 \end{bmatrix}$.

5. Not in echelon form.
$R_1 \leftrightarrow R_2$, $(1/2)R_1$, $(1/2)R_2$ yields the echelon form $\begin{bmatrix} 1 & 0 & 1/2 & 2 \\ 0 & 0 & 1 & 3/2 \end{bmatrix}$.

7. Not in echelon form. $R_2 - 4R_3$, $R_1 - 2R_3$, $R_1 - 3R_2$ yields the reduced echelon form $\begin{bmatrix} 1 & 0 & 0 & 5 \\ 0 & 1 & 0 & -2 \\ 0 & 0 & 1 & 1 \end{bmatrix}$.

9. Not in echelon form. $(1/2)R_2$ yields the echelon form $\begin{bmatrix} 1 & 2 & -1 & -2 \\ 0 & 1 & -1 & -3/2 \\ 0 & 0 & 0 & 1 \end{bmatrix}$.

11. $x_1 = 0$, $x_2 = 0$.

13. $x_1 = -2 + 5x_3$, $x_2 = 1 - 3x_3$, x_3 is arbitrary.

15. $x_1 = 0$, $x_2 = 0$, $x_3 = 0$.

17. $x_1 = x_3 = x_4 = 0$, x_2 is arbitrary.

19. The system is inconsistent.

21. $x_1 = -1 - (1/2)x_2 + (1/2)x_4$, $x_3 = 1 - x_4$, x_2 and x_4 arbitrary, $x_5 = 0$.

23. The system is inconsistent.

25. $x_1 = 2 - x_2$, x_2 arbitrary.

27. $x_1 = 2 - x_2 + x_3$, x_2 and x_3 arbitrary.

29. $x_1 = 3 - 2x_3$, $x_2 = -2 + 3x_3$, x_3 arbitrary.

31. $x_1 = 3 - (7x_4 - 16x_5)/2$, $x_2 = (x_4 + 2x_5)/2$, $x_3 = -2 + (5x_4 - 12x_5)/2$, x_4 and x_5 arbitrary.

33. The system is inconsistent.

35. The system is inconsistent.

37. $$\begin{array}{rcl} x_1 + 3x_2 & = & 4 \\ 2x_1 + 6x_2 & = & a \end{array} \left\{ \begin{array}{c} E_2 - 2E_1 \\ \Longrightarrow \end{array} \right\} \begin{array}{rcl} x_1 + 3x_2 & = & 4 \\ 0 & = & a - 8 \end{array}$$

Thus, if $a \neq 8$ there is no solution.

39. $$\begin{array}{rcl} 3x_1 + ax_2 & = & 3 \\ ax_1 + 3x_2 & = & 5 \end{array} \left\{ \begin{array}{c} E_2 - (a/3)E_1 \\ \Longrightarrow \end{array} \right\} \begin{array}{rcl} 3x_1 + ax_2 & = & 3 \\ (a^2/3 - 3)x_2 & = & 5 - a \end{array}$$

Thus, if $a = \pm 3$ there is no solution.

41. $\cos\alpha = 1/2$ and $\sin\beta = 1/2$, so $\alpha = \pi/3$ or $\alpha = 5\pi/3$ and $\beta = \pi/6$ or $\beta = 5\pi/6$.

43. $x_1 = 1 - 2x_3$, $x_2 = 2 + x_3, x_3$ arbitrary. (a) $x_3 = 1/2$. (b) In order for $x_1 \geq 0, x_2 \geq 0$, we must have $-2 \leq x_3 \leq 1/2$; for a given x_1 and $x_2, y = -6 - 7x_3$, so the minimum value is $y = 8$ at $x_3 = -2$. (c) The minimum value is 20.

45. $$\begin{bmatrix} 1 & x & x \\ 0 & 1 & x \end{bmatrix}, \begin{bmatrix} 1 & x & x \\ 0 & 0 & 1 \end{bmatrix}, \begin{bmatrix} 1 & x & x \\ 0 & 0 & 0 \end{bmatrix}, \begin{bmatrix} 0 & 1 & x \\ 0 & 0 & 1 \end{bmatrix}$$
$$\begin{bmatrix} 0 & 1 & x \\ 0 & 0 & 0 \end{bmatrix}, \begin{bmatrix} 0 & 0 & 1 \\ 0 & 0 & 0 \end{bmatrix}, \begin{bmatrix} 0 & 0 & 0 \\ 0 & 0 & 0 \end{bmatrix}.$$

47. $$\begin{bmatrix} 1 & 2 \\ 2 & 3 \end{bmatrix} \left\{ \begin{array}{c} 2R_2 \\ \Longrightarrow \end{array} \right\} \begin{bmatrix} 1 & 2 \\ 4 & 6 \end{bmatrix} \left\{ \begin{array}{c} R_2 - R_1 \\ \Longrightarrow \end{array} \right\} \begin{bmatrix} 1 & 2 \\ 3 & 4 \end{bmatrix}.$$

49. $$\begin{array}{rcl} 100x_1 + 10x_2 + x_3 & = & 15(x_1 + x_2 + x_3) \\ 100x_3 + 10x_2 + x_1 & = & 100x_1 + 10x_2 + x_3 + 396 \\ x_3 & = & x_1 + x_2 + 1 \end{array}$$ $x_1 = 1$, $x_2 = 3$, and $x_3 = 5$, so $N = 135$.

51. Let x_1, x_2, x_3 be the amounts initially held by players one, two and three, respectively. Also assume that player one loses the first game, player two loses the second game, and player three loses the third game. Then after three games, the amount of money held by each player is given by the following table

Player	**Amount of money**
1	$4x_1 - 4x_2 - 4x_3 = 24$
2	$-2x_1 + 6x_2 - 2x_3 = 24$
3	$-x_1 - x_2 + 7x_3 = 24$

Solving yields $x_1 = 39$, $x_2 = 21$, and $x_3 = 12$.

53. If x_1 is the number of adults, x_2 the number of students, and x_3 the number of children, then $x_1 + x_2 + x_3 = 79$, $6x_1 + 3x_2 + (1/2)x_3 = 207$, and for $j = 1, 2, 3$, x_j is an integer such that $0 \le x_j \le 79$. Following is a list of possiblities

Number of Adults	0	5	10	15	20	25	30
Number of Students	67	56	45	34	23	12	1
Number of Children	12	18	24	30	36	42	48

55. By (7), $1 + 2 + 3 + \cdots + n = a_1 n + a_2 n^2$. Setting $n = 1$ and $n = 2$ gives

$$\begin{aligned} a_1 + a_2 &= 1 \\ 2a_1 + 4a_2 &= 3 \end{aligned}$$

The solution is $a_1 = a_2 = 1/2$, so $1 + 2 + 3 + \ldots + n = n(n+1)/2$.

57. The system of equations obtained from (7) is

$$\begin{aligned} a_1 + a_2 + a_3 + a_4 + a_5 &= 1 \\ 2a_1 + 4a_2 + 8a_3 + 16a_4 + 32a_5 &= 17 \\ 3a_1 + 9a_2 + 27a_3 + 81a_4 + 242a_5 &= 98 \\ 4a_1 + 16a_2 + 64a_3 + 256a_4 + 1024a_5 &= 354 \\ 5a_1 + 25a_2 + 125a_3 + 625a_4 + 3125a_5 &= 979 \end{aligned}$$

The solution is $a_1 = -1/30$, $a_2 = 0$, $a_3 = 1/3$, $a_4 = 1/2$, $a_5 = 1/5$. Therefore, $1^4 + 2^4 + 3^4 + \cdots + n^4 = n(n+1)(2n+1)(3n^2 + 3n - 1)/30$.

1.3 Consistent Systems of Linear Equations

1. The augmented matrix reduces to $\begin{bmatrix} 1 & 1 & 0 & 5/6 \\ 0 & 0 & 1 & 2/3 \\ 0 & 0 & 0 & 0 \\ 0 & 0 & 0 & 0 \end{bmatrix}$.

   n = 3, r = 2, x_2 is independent.

3. The augmented matrix reduces to $\begin{bmatrix} 1 & 0 & 4 & 0 & 13/2 \\ 0 & 1 & -1 & 0 & -3/2 \\ 0 & 0 & 0 & 1 & 1/2 \end{bmatrix}$.

   n = 4, r = 3, x_3 is independent.

5. $n = 2$ and $r \le 2$ so $r = 0$, $n - r = 2$; $r = 1$, $n - r = 1$; $r = 2$. $n - r = 0$. There could be a unique solution.

7. Infinitely many solutions or no solution.

9. Infinitely many solutions.

11. Infinitely many solutions, a unique solution or no solution.

13. A unique solution or infinitely many solutions.

15. Infinitely many solutions.

17. Infinitely many solutions or a unique solution.

19. Infinitely many solutions.

21. There are nontrivial solutions.

23. There is only the trivial solution.

25. (a) $B = \begin{bmatrix} * & x & x \\ 0 & * & x \\ 0 & 0 & * \\ 0 & 0 & 0 \end{bmatrix}$.

(b) In the third row of the matrix of 25(a) for B, we need $0 \cdot x_1 + 0 \cdot x_2 = *$ and, in general, this can't be.

27. The resulting system of equations is $\begin{aligned} 2a + 8b + c &= 0 \\ 4a + b + c &= 0 \end{aligned}$.
The general solution is a = (-7/30)c, b = (-1/15)c.
Thus $-7x - 2y + 30 \;=\; 0$ is an equation for the line.

29. The resulting system of equations is
$$\begin{aligned} 16a - 4b + c - 4d + e + f &= 0 \\ a - 2b + 4c - d + 2e + f &= 0 \\ 9a + 6b + 4c + 3d + 2e + f &= 0\,. \\ 25a + 5b + c + 5d + e + f &= 0 \\ 49a - 7b + c + 7d - e + f &= 0 \end{aligned}$$
The general solution is:
a = (-3/113)f, b = (3/113)f, c = (1/113)f, d = 0, e = (-54/113)f.
An equation is $-3x^2 + 3xy + y^2 - 54y + 113 \;=\; 0$.

31. Omitted

33. The resulting system of equations is: $\begin{aligned} 25a + 4b + 3c + d &= 0 \\ 5a + b + 2c + d &= 0\,. \\ 4a + 2b + d &= 0 \end{aligned}$
The general solution is: $a = (7/50)d, b = (-39/50)d, c = (-23/50)d$.
Thus, $7x^2 + 7y^2 - 39x - 23y + 50 \;=\; 0$,
is an equation for the circle.

1.4 Applications

1. (a)
$$\begin{aligned} x_1 + x_4 &= 1200 \\ x_1 + x_2 &= 1000 \\ x_3 + x_4 &= 600 \\ x_2 + x_4 &= 400 \end{aligned}$$
The solution is $x_1 = 1200 - x_4$, $x_2 = -200 + x_4$, $x_3 = 600 - x_4$.

   (b) $x_1 = 1100$, $x_2 = -100$, $x_3 = 500$.

   (c) $200 \leq x_4 \leq 600$ so $600 \leq x_1 \leq 1000$

3. $x_2 = 800$, $x_3 = 400$, $x_4 = 200$.

5. $4I_1 + 3I_2 = 2$, $3I_2 + 4I_3 = 4$, and $I_1 + I_3 = I_2$. Therefore, $I_1 = 1/20$, $I_2 = 3/5$, and $I_3 = 11/20$.

$5/7, 20/7, 15/7$

9. (a)
$$\begin{aligned} x_1 - x_4 &= a_1 - a_2 \\ x_1 - x_2 &= -b_1 + b_2 \\ -x_3 + x_4 &= d_1 - d_2 \\ x_2 - x_3 &= -c_1 + c_2 \end{aligned}$$

1.5 Matrix Operations

1. (a) $\begin{bmatrix} 2 & 0 \\ 2 & 6 \end{bmatrix}$, (b) $\begin{bmatrix} 0 & 4 \\ 2 & 4 \end{bmatrix}$, (c) $\begin{bmatrix} 0 & -6 \\ 6 & 18 \end{bmatrix}$, (d) $\begin{bmatrix} -6 & 8 \\ 4 & 6 \end{bmatrix}$.

3. $\begin{bmatrix} -2 & -2 \\ 0 & 0 \end{bmatrix}$.

5. $\begin{bmatrix} -1 & -1 \\ 0 & 0 \end{bmatrix}$.

7. (a) $\begin{bmatrix} 3 \\ -3 \end{bmatrix}$, (b) $\begin{bmatrix} 3 \\ 4 \end{bmatrix}$, (c) $\begin{bmatrix} 0 \\ 0 \end{bmatrix}$.

9. (a) $\begin{bmatrix} 2 \\ 1 \end{bmatrix}$, (b) $\begin{bmatrix} 0 \\ 1 \end{bmatrix}$, (c) $\begin{bmatrix} 17 \\ 14 \end{bmatrix}$.

11. (a) $\begin{bmatrix} 2 \\ 3 \end{bmatrix}$, (b) $\begin{bmatrix} 20 \\ 16 \end{bmatrix}$.

13. $a_1 = 11/3$, $a_2 = -(4/3)$.

15. $a_1 = -2$, $a_2 = 0$.

17. The equation has no solution.

19. $a_1 = 4$, $a_2 = -(3/2)$.

21. $\mathbf{w}_1 = \begin{bmatrix} 0 \\ 1 \end{bmatrix}$, $\mathbf{w}_2 = \begin{bmatrix} 1 \\ 3 \end{bmatrix}$, $AB = \begin{bmatrix} 1 & 1 \\ 3 & 8 \end{bmatrix}$, $(AB)\mathbf{r} = \begin{bmatrix} 1 \\ 3 \end{bmatrix}$

23. $\mathbf{w}_1 = \begin{bmatrix} -2 \\ 1 \end{bmatrix}$, $\mathbf{w}_2 = \begin{bmatrix} -1 \\ 1 \end{bmatrix}$, $\mathbf{w}_3 = \begin{bmatrix} -1 \\ 2 \end{bmatrix}$, $Q = \begin{bmatrix} -1 & 4 \\ 2 & 17 \end{bmatrix}$,

$Q\,\mathbf{r} = \begin{bmatrix} -1 \\ 2 \end{bmatrix}$.

25. $\begin{bmatrix} -4 & 6 \\ 2 & 12 \end{bmatrix}$.

27. $\begin{bmatrix} 4 & 12 \\ 4 & 10 \end{bmatrix}$.

29. $\begin{bmatrix} 0 & 0 \\ 0 & 0 \end{bmatrix}$.

31. $AB = \begin{bmatrix} 5 & 16 \\ 5 & 18 \end{bmatrix}$, $BA = \begin{bmatrix} 4 & 11 \\ 6 & 19 \end{bmatrix}$.

33. $A\,\mathbf{u} = \begin{bmatrix} 11 \\ 13 \end{bmatrix}$, $\mathbf{v}\,A = [8, 22]$.

35. $\mathbf{v}\,B\mathbf{u} = 66$.

37. $CA = \begin{bmatrix} 5 & 10 \\ 8 & 12 \\ 15 & 20 \\ 8 & 17 \end{bmatrix}$.

39. $C(B)\,\mathbf{u}) = \begin{bmatrix} 27 \\ 28 \\ 43 \\ 47 \end{bmatrix}$.

41. $(BA)\mathbf{u} = \begin{bmatrix} 37 \\ 63 \end{bmatrix}$, $B(A\mathbf{u}) = \begin{bmatrix} 37 \\ 63 \end{bmatrix}$.

43. $\begin{bmatrix} x_1 \\ x_2 \\ x_3 \end{bmatrix} = \begin{bmatrix} -2+x_3 \\ 3-2x_3 \\ x_3 \end{bmatrix} = \begin{bmatrix} -2 \\ 3 \\ 0 \end{bmatrix} + x_3 \begin{bmatrix} 1 \\ -2 \\ 1 \end{bmatrix}.$

45. $\begin{bmatrix} x_1 \\ x_2 \\ x_3 \\ x_4 \\ x_5 \end{bmatrix} = \begin{bmatrix} x_3+x_5 \\ -2x_3-x_5 \\ x_3 \\ -x_5 \\ x_5 \end{bmatrix} = x_3 \begin{bmatrix} 1 \\ -2 \\ 1 \\ 0 \\ 0 \end{bmatrix} + x_5 \begin{bmatrix} 1 \\ -1 \\ 0 \\ -1 \\ 1 \end{bmatrix}.$

47. $\begin{bmatrix} x_1 \\ x_2 \\ x_3 \\ x_4 \\ x_5 \end{bmatrix} = \begin{bmatrix} x_3+2x_4+3x_5 \\ -2x_3-3x_4-4x_5 \\ x_3 \\ x_4 \\ x_5 \end{bmatrix} = x_3 \begin{bmatrix} 1 \\ -2 \\ 1 \\ 0 \\ 0 \end{bmatrix} + x_4 \begin{bmatrix} 2 \\ -3 \\ 0 \\ 1 \\ 0 \end{bmatrix} + x_5 \begin{bmatrix} 3 \\ -4 \\ 0 \\ 0 \\ 1 \end{bmatrix}.$

49. $\begin{bmatrix} x_1 \\ x_2 \\ x_3 \\ x_4 \\ x_5 \end{bmatrix} = \begin{bmatrix} x_2+2x_4 \\ x_2 \\ -2x_4 \\ x_4 \\ 0 \end{bmatrix} = x_2 \begin{bmatrix} 1 \\ 1 \\ 0 \\ 0 \\ 0 \end{bmatrix} + x_4 \begin{bmatrix} 2 \\ 0 \\ -2 \\ 1 \\ 0 \end{bmatrix}.$

51. $C(A(B\mathbf{u}))$ has 12 multiplications, $(CA)(B\mathbf{u})$ has 16 multiplications, $[C(AB)](\mathbf{u})$ has 20 multiplications, and $C[(AB)\mathbf{u}]$ has 16 multiplications.

53. (a) AB is a 2 x 4 matrix, BA is not defined.

(b) AB is not defined, BA is not defined.

(c) AB is not defined. BA is a 6 x 7 matrix.

(d) AB is a 2 x 2 matrix, BA is a 3 x 3 matrix.

(e) AB is a 3 x 1 matrix, BA is not defined.

(f) $A(BC)$ and $(AB)C$ are 2 x 4 matrices.

(g) AB is a 4 x 4 matrix. BA is a 1 x 1 matrix.

55. $A^2 = AA$ provided A is a square matrix.

57. (a) $P\mathbf{x} = \begin{bmatrix} 135,000 \\ 120,000 \\ 45,000 \end{bmatrix}$ is the state vector after one year and $P^2\mathbf{x} = \begin{bmatrix} 126,000 \\ 132,000 \\ 42,000 \end{bmatrix}$ is the state vectore after two years.

(b) $P^n\mathbf{x}$

59. Let A be an $(m \times n)$ matrix and B be a $(p \times r)$ matrix. Since AB is defined, $n = p$ and AB is an $(m \times r)$ matrix. But AB is a square matrix, so $m = r$. Thus, B is an $(n \times m)$ matrix, so BA is defined and is an $(n \times n)$ matrix.

61. (a) (i) $A = \begin{bmatrix} 2 & -1 \\ 1 & 1 \end{bmatrix}, \mathbf{x} = \begin{bmatrix} x_1 \\ x_2 \end{bmatrix}, \mathbf{b} = \begin{bmatrix} 3 \\ 3 \end{bmatrix}.$

(ii) $A = \begin{bmatrix} 1 & -3 & 1 \\ 1 & -2 & 1 \\ 0 & 1 & -1 \end{bmatrix}, \mathbf{x} = \begin{bmatrix} x_1 \\ x_2 \\ x_3 \end{bmatrix}, \mathbf{b} = \begin{bmatrix} 1 \\ 2 \\ -1 \end{bmatrix}.$

(b) (i) $x_1 \begin{bmatrix} 2 \\ 1 \end{bmatrix} + x_2 \begin{bmatrix} -1 \\ 1 \end{bmatrix} = \begin{bmatrix} 3 \\ 3 \end{bmatrix}.$

(ii) $x_1 \begin{bmatrix} 1 \\ 1 \\ 0 \end{bmatrix} + x_2 \begin{bmatrix} -3 \\ -2 \\ 1 \end{bmatrix} + x_3 \begin{bmatrix} 1 \\ 1 \\ -1 \end{bmatrix} = \begin{bmatrix} 1 \\ 2 \\ -1 \end{bmatrix}.$

(c) (i) $x_1 = 2,\ x_2 = 1,\ 2\mathbf{A_1} + \mathbf{A_2} = \mathbf{b}.$

(ii) $x_1 = 2,\ x_2 = 1,\ x_3 = 2,\ 2\mathbf{A_1} + \mathbf{A_2} + 2\mathbf{A_3} = \mathbf{b}.$

63. (a) We solve each of the systems

(i) $A\mathbf{x} = \begin{bmatrix} 1 \\ 0 \end{bmatrix},$

(ii)) $A\mathbf{x} = \begin{bmatrix} 0 \\ 1 \end{bmatrix}.$

(i) $\mathbf{x} = \begin{bmatrix} 2 \\ -1 \end{bmatrix}$; (ii) $\mathbf{x} = \begin{bmatrix} -1 \\ 1 \end{bmatrix}.$

(b) $B = \begin{bmatrix} 2 & -1 \\ -1 & 1 \end{bmatrix}$ and$AB = I = BA.$

65. (a) $B = \begin{bmatrix} -1 & 6 \\ 1 & 0 \end{bmatrix}$. (b) No B exists. (c) $B = \begin{bmatrix} 2 & 2 \\ -1 & -1 \end{bmatrix}.$

69. $$\begin{bmatrix} x_1 \\ x_2 \\ x_3 \\ x_4 \\ x_5 \end{bmatrix} = \begin{bmatrix} 5 + 2x_4 - 3x_5 \\ 4 - 3x_4 - 2x_5 \\ 2 - x_4 - x_5 \\ x_4 \\ x_5 \end{bmatrix} = \begin{bmatrix} 5 \\ 4 \\ 2 \\ 0 \\ 0 \end{bmatrix} + x_4 \begin{bmatrix} 2 \\ -3 \\ -1 \\ 1 \\ 0 \end{bmatrix} + x_5 \begin{bmatrix} -3 \\ -2 \\ -1 \\ 0 \\ 1 \end{bmatrix}.$$

1.6 Algebraic Properties of Matrix Operations

1. $DE = \begin{bmatrix} 8 & 15 \\ 11 & 18 \end{bmatrix}$, $EF = \begin{bmatrix} 9 & 9 \\ 5 & 5 \end{bmatrix}$, $(DE)F = D(EF) =$

$\begin{bmatrix} 23 & 23 \\ 29 & 29 \end{bmatrix}$.

3. $DE = \begin{bmatrix} 8 & 15 \\ 11 & 18 \end{bmatrix}$, $ED = \begin{bmatrix} 12 & 27 \\ 7 & 14 \end{bmatrix}$.

5. $F\mathbf{u} = \begin{bmatrix} 0 \\ 0 \end{bmatrix}$, $F\mathbf{v} = \begin{bmatrix} 0 \\ 0 \end{bmatrix}$.

7. $A^{\mathrm{T}} = \begin{bmatrix} 3 & 4 & 2 \\ 1 & 7 & 6 \end{bmatrix}$.

9. $E^{\mathrm{T}}F = \begin{bmatrix} 5 & 5 \\ 9 & 9 \end{bmatrix}$.

11. $(F\mathbf{v})^{\mathrm{T}} = \begin{bmatrix} 0 & 0 \end{bmatrix}$.

13. -6.

15. 36.

17. 2.

19. $\sqrt{2}$.

21. $\sqrt{29}$.

23. 0.

25. $2\sqrt{5}$

27. Let $A = \begin{bmatrix} 1 & 0 \\ 0 & 0 \end{bmatrix}$ and let $B = \begin{bmatrix} 1 & 0 \\ 0 & 1 \end{bmatrix}$. Then $A^2 = AB$ and $A \neq B$.

29. D and F are symmetric.

31. If each of A and B are symmetric, then a necessary and sufficient condition that AB be symmetric is that $AB = BA$.

33. $\mathbf{x}^{\mathrm{T}} D\mathbf{x} = [x_2, x_2] \begin{bmatrix} 2 & 2 \\ 1 & 4 \end{bmatrix} \begin{bmatrix} x_1 \\ x_2 \end{bmatrix} = x_1^2 + 3x_2^2 + (x_1 + x_2)^2$. This term is always greater than zero whenever x_1 and x_2 are not simultaneously zero.

35. $\begin{bmatrix} -3 & 3 \\ 3 & -3 \end{bmatrix}$.

37. $\begin{bmatrix} -27 & -9 \\ 27 & 9 \end{bmatrix}$.

39. $\begin{bmatrix} -12 & 18 & 24 \\ 18 & -27 & -36 \\ 24 & 36 & -48 \end{bmatrix}$.

41. (a) $\mathbf{x}^T\mathbf{a} = 6$ means that $x_1 + 2x_2 = 6$ and $\mathbf{x}^T\mathbf{b} = 2$ means that $3x_1 + 4x_2 = 2$. Thus $x_1 = -10$, $x_2 = 8$ and $\mathbf{x} = \begin{bmatrix} -10 \\ 8 \end{bmatrix}$.

(b) $\mathbf{x}^T(\mathbf{a}+\mathbf{b}) = 12$ and $\mathbf{x}^T\mathbf{a} = 2$ yields $4x_1 + 6x_2 = 12$ and $x_1 + 2x_2 = 2$. Thus $x_1 = 6$, $x_2 = -2$ and $\mathbf{x} = \begin{bmatrix} 6 \\ -2 \end{bmatrix}$.

57. n = 5, m = 7.

59. n = 4, m = 6.

61. n = m = 5.

1.7 Linear Independence and Nonsingular Matrices

1. $x_1\mathbf{v}_1 + x_2\mathbf{v}_2 = \theta$ has only the trivial solution so $\{\mathbf{v}_1, \mathbf{v}_2\}$ is linearly independent.

3. Linearly dependent. $\mathbf{v}_5 = 3\mathbf{v}_1$.

5. Linearly dependent. $\mathbf{v}_3 = 2\mathbf{v}_1$.

7. Linearly dependent. $\mathbf{u}_4 = 4\mathbf{u}_5$.

9. $x_1\mathbf{u}_1 + x_2\mathbf{u}_2 + x_3\mathbf{u}_5 = \theta$ has only the trivial solution so $\{\mathbf{u}_1, \mathbf{u}_2, \mathbf{u}_5\}$ is linearly independent.

11. Linearly dependent. $\mathbf{u}_4 = 4\mathbf{u}_5$.

13. Linearly dependent. $\mathbf{u}_4 = (16/5)\mathbf{u}_0 + (12/5)\mathbf{u}_1 - (4/5)\mathbf{u}_2$.

15. Sets 5, 6, 13, and 14 are linearly dependent by inspection.

17. B is singular, $x_1 = -2x_2$.

19. AB is singular, $x_1 = -2x_2$.

21. D is singular, $x_1 = x_2 = 0$, x_3 arbitrary.

23. $D + F$ is nonsingular.

25. EF is singular, x_1 arbitrary, $x_2 = 0 = x_3$.

27. F^{T} is nonsingular.

29. $a = 6$.

31. $\{\mathbf{v}_1, \mathbf{v}_2, \mathbf{v}_3\}$ is linearly dependent if $b(a-2) = 4$.

33. $\{\mathbf{v}_1, \mathbf{v}_2\}$ is linearly dependent if $c = ab$.

35. $\mathbf{x} = \begin{bmatrix} 0 \\ 1 \end{bmatrix}$, $\mathbf{v}_3 = \mathbf{A}_2$.

37. $\mathbf{x} = \begin{bmatrix} 1/2 \\ 1/2 \end{bmatrix}$, $\mathbf{v}_2 = (1/2)(\mathbf{C}_1 + \mathbf{C}_2)$.

39. $\mathbf{x} = \begin{bmatrix} -8/3 \\ -2/3 \\ 3 \end{bmatrix}$ $\mathbf{u}_3 = (-8\mathbf{F}_1 - 2\mathbf{F}_2 + 9\mathbf{F}_3)/3$.

41. $-11\begin{bmatrix} 1 \\ 2 \end{bmatrix} + 7\begin{bmatrix} 2 \\ 3 \end{bmatrix} = \begin{bmatrix} 3 \\ -1 \end{bmatrix}$.

43. $0\begin{bmatrix} 1 \\ 2 \end{bmatrix} + 0\begin{bmatrix} 2 \\ 3 \end{bmatrix} = \begin{bmatrix} 0 \\ 0 \end{bmatrix}$.

45. $-3\begin{bmatrix} 1 \\ 2 \end{bmatrix} + 2\begin{bmatrix} 2 \\ 3 \end{bmatrix} = \begin{bmatrix} 1 \\ 0 \end{bmatrix}$.

1.8 Data Fitting, Numerical Integration & Differentiation

1. $p(t) = (-1/2)t^2 + (9/2)t - 1$.

3. $p(t) = 2t + 3$.

5. $p(t) = 2t^3 - 2t^2 + 3t + 1$.

7. $y = 2e^{2x} + e^{3x}$.

9. $y = 3e^{-x} + 4e^{x} + e^{2x}$.

11. $\int_0^{3h} f(t)dt \approx \frac{3h}{2}[f(h) + f(2h)]$.

13. $\int_0^{3h} f(t)dt \approx \frac{3h}{8}[f(0) + 3f(h) + 3f(2h) + f(3h)]$.

15. $\int_0^{h} f(t)dt \approx \frac{h}{2}[-f(-h) + 3f(0)]$.

17. $f'(0) \approx \frac{-1}{h}f(0) + \frac{1}{h}f(h)$.

19. $f'(0) \approx -\frac{3}{2h}f(0) + \frac{2}{h}f(h) + \frac{-1}{2h}f(2h)$.

21. $f''(0) \approx \frac{1}{h^2}[f(-h) - 2f(0) + f(h)]$.

23. $\begin{bmatrix} 1 & 1 & 1 & b-a \\ a & t & b & (b^2-a^2)/2 \\ a^2 & t^2 & b^2 & (b^3-a^3)/3 \end{bmatrix}$

$$\begin{bmatrix} 1 & 1 & 1 & b-a \\ 0 & t-a & b-a & (b^2+a^2-2ab)/2 \\ 0 & t^2-a^2 & b^2-a^2 & (b^3-a^3-3(b-a)a^2)/3 \end{bmatrix}$$

$$\begin{bmatrix} 1 & 1 & 1 & b-a \\ 0 & 1 & 2 & b-a \\ 0 & t^2-a^2 & b^2-a^2 & (b^3-a^3-3(b-a)a^2)/3 \end{bmatrix}$$

$$\begin{bmatrix} 1 & 1 & 1 & b-a \\ 0 & 1 & 2 & b-a \\ 0 & 0 & (b-a)^2/2 & (b-a)^3/12 \end{bmatrix} \begin{bmatrix} 1 & 1 & 1 & b-a \\ 0 & 1 & 2 & b-a \\ 0 & 0 & 6 & b-a \end{bmatrix}$$

27. We must solve the system $L\mathbf{x} = \mathbf{b}$ where $L = \begin{bmatrix} 0 & 0 & 0 & 1 \\ 0 & 0 & 1 & 0 \\ 1 & 1 & 1 & 1 \\ 3 & 2 & 1 & 0 \end{bmatrix}$

$$\mathbf{x} = \begin{bmatrix} a \\ b \\ c \\ d \end{bmatrix}, \mathbf{b} = \begin{bmatrix} 2 \\ 3 \\ 8 \\ 10 \end{bmatrix}. \; p(t) = t^3 + 2t^2 + 3t + 2.$$

29. $p(t) = t^3 + t^2 + 4t + 3$.

35. $f'(a) \approx \frac{1}{12h}[f(a-2h) - 8f(a-h) + 8f(a+h) - f(a+2h)]$

1.9 Matrix Inverses and their Properties

5. (cf. Ex. 2) $\mathbf{x} = A^{-1}\mathbf{b} = B\mathbf{b} = \begin{bmatrix} 1 & -1 \\ -.2 & .3 \end{bmatrix} \begin{bmatrix} 6 \\ 9 \end{bmatrix} = \begin{bmatrix} -3 \\ 1.5 \end{bmatrix}$

7. (cf. Ex. 3) $\mathbf{x} = B^{-1}\mathbf{b} = A\mathbf{b} = \begin{bmatrix} -1 & -2 & 11 \\ 1 & 3 & -15 \\ 0 & -1 & 5 \end{bmatrix} \begin{bmatrix} 4 \\ 2 \\ 2 \end{bmatrix} = \begin{bmatrix} 14 \\ -20 \\ 8 \end{bmatrix}$.

9. If B is any 3 x 3 matrix, then the $(1,1)^{th}$ entry of AB is zero and so $AB \neq I$.

11. Let $B = (x_{ij})$ be a (3 x 3) matrix and suppose that $BA = I$. Then the $(1,1)^{th}$ entry of BA must be one and the $(1,2)^{th}$ entry of BA be must be zero. But each of these entries equals $2x_{11} + x_{12} + 3x_{13}$ and cannot simultaneously be one and zero.

13. $\begin{bmatrix} 3 & -1 \\ -2 & 1 \end{bmatrix}$.

15. $\begin{bmatrix} -1/3 & 2/3 \\ 2/3 & -1/3 \end{bmatrix}$.

17. $\begin{bmatrix} 1 & 0 & 0 \\ -2 & 1 & 0 \\ 5 & -4 & 1 \end{bmatrix}$.

19. $\begin{bmatrix} 1 & -2 & 0 \\ 3 & -3 & -1 \\ -6 & 7 & 2 \end{bmatrix}$.

21. $\begin{bmatrix} -1/2 & -2/3 & -1/6 & 7/6 \\ 1 & 1/3 & 1/3 & -4/3 \\ 0 & -1/3 & -1/3 & 1/3 \\ -1/2 & 1 & 1/2 & 1/2 \end{bmatrix}$.

23. $\Delta = 10$ so $A^{-1} = \frac{1}{10}\begin{bmatrix} 3 & 2 \\ -2 & 2 \end{bmatrix}$.

25. $\Delta = 0$ so A^{-1} does not exist.

27. $\mathbf{x} = A^{-1}\mathbf{b} = \begin{bmatrix} 2 & -1 \\ -3 & 2 \end{bmatrix} \begin{bmatrix} 4 \\ 2 \end{bmatrix} = \begin{bmatrix} 6 \\ -8 \end{bmatrix}$.

29. $\mathbf{x} = A^{-1}\mathbf{b} = \begin{bmatrix} 4 & -1 \\ 3 & -1 \end{bmatrix} \begin{bmatrix} 5 \\ 2 \end{bmatrix} = \begin{bmatrix} 18 \\ 13 \end{bmatrix}$.

31. $\mathbf{x} = A^{-1}\mathbf{b} = \frac{1}{10}\begin{bmatrix} 3 & -1 \\ 1 & 3 \end{bmatrix} \begin{bmatrix} 10 \\ 5 \end{bmatrix} = \begin{bmatrix} 5/2 \\ 5/2 \end{bmatrix}$.

33. $Q^{-1} = C^{-1}A^{-1} = \begin{bmatrix} -3 & 1 \\ 3 & 5 \end{bmatrix}$.

35. $Q^{-1} = (A^{-1})^{\mathrm{T}} = \begin{bmatrix} 3 & 0 \\ 1 & 2 \end{bmatrix}$.

37. $Q^{-1} = (A^{-1})^{\mathrm{T}}(C^{-1})^{\mathrm{T}} = \begin{bmatrix} 3 & 0 \\ 1 & 2 \end{bmatrix}\begin{bmatrix} -1 & 1 \\ 1 & 2 \end{bmatrix} = \begin{bmatrix} -3 & 3 \\ 1 & 5 \end{bmatrix}$.

39. $Q^{-1} = BC^{-1} = \begin{bmatrix} 1 & 5 \\ -1 & 4 \end{bmatrix}$.

41. $Q^{-1} = \frac{1}{2}A^{-1} = \begin{bmatrix} 3/2 & 1/2 \\ 0 & 1 \end{bmatrix}$.

43. $Q^{-1} = B(C^{-1}A^{-1}) = \begin{bmatrix} 3 & 11 \\ -3 & 7 \end{bmatrix}$.

45. $B = A^{-1}D = \begin{bmatrix} 1 & 10 \\ 15 & 12 \\ 3 & 3 \end{bmatrix}, C = EA^{-1}, = \begin{bmatrix} 13 & 12 & 8 \\ 2 & 3 & 5 \end{bmatrix}$.

47. $(AB)^{-1} = B^{-1}A^{-1} = \begin{bmatrix} 2 & 35 & 1 \\ 14 & 35 & 34 \\ 23 & 12 & 70 \end{bmatrix}$

$(3A)^{-1} = (1/3)A^{-1} = \begin{bmatrix} 1/3 & 2/3 & 5/3 \\ 1 & 1/3 & 2 \\ 2/3 & 8/3 & 1/3 \end{bmatrix}$

$(A^T)^{-1} = (A^{-1})^T = \begin{bmatrix} 1 & 3 & 2 \\ 2 & 1 & 8 \\ 5 & 6 & 1 \end{bmatrix}$

61. $\theta = A^2 + b_1A + b_0I = \begin{bmatrix} 11 & -4 \\ -2 & 3 \end{bmatrix} + \begin{bmatrix} -3b_1 & 2b_1 \\ b_1 & b_1 \end{bmatrix} + \begin{bmatrix} b_0 & 0 \\ 0 & b_0 \end{bmatrix}$

$= \begin{bmatrix} 11 - 3b_1 + b_0 & -4 + 2b_1 \\ -2 + b_1 & 3 + b_1 + b_0 \end{bmatrix}.\mathrm{b}_0 = -5, \mathrm{b}_1 = 2.$

$(-1/\mathrm{b}_0)[A + b_1 I] = -(1/5)\begin{bmatrix} 1 & -2 \\ -1 & -3 \end{bmatrix} = A^{-1}.$

63. $\theta = A^2 + b_1A + b_0I = \begin{bmatrix} 7 & 0 \\ 0 & 7 \end{bmatrix} + \begin{bmatrix} -b_1 & +3b_1 \\ 2b_1 & b_1 \end{bmatrix} + \begin{bmatrix} b_0 & 0 \\ 0 & b_0 \end{bmatrix} = \begin{bmatrix} 7 - b_1 + b_0 & 3b_1 \\ 2b_1 & 7 + b_1 + b_0 \end{bmatrix}$.

$\mathrm{b}_0 = -7, \mathrm{b}_1 = 0$

$(-1/b_0)[A + b_1I] = \frac{1}{7}\begin{bmatrix} -1 & 3 \\ 2 & 1 \end{bmatrix} = A^{-1}.$

65. (a) $A = \begin{bmatrix} 1 & 0 \\ 0 & 1 \end{bmatrix}$ and $B = \begin{bmatrix} 0 & 1 \\ 1 & 0 \end{bmatrix}$.

(b) $A = \begin{bmatrix} 1 & 0 \\ 0 & 0 \end{bmatrix}$ and $B = \begin{bmatrix} 0 & 0 \\ 0 & 1 \end{bmatrix}$.

Chapter 2

The Vector Space R^n

2.1 Introduction

13. Geometrically, W consists of the points in the plane that lie on the line with equation $x = -3y$.

15. Geometrically, W consists of the points in the plane that have coordinates (x, y) satisfying $x + y \geq 0$.

17. Geometrically, W consists of the points in the plane that lie on the circle with equation $x^2 + y^2 = 4$.

19. Geometrically, W consists of the points in three-space that lie on the plane with equation $x + y + 2z = 0$.

21. Geometrically, W consists of the points in three-space that are on or above the xy plane and that lie on the sphere with equation $x^2 + y^2 + z^2 = 1$.

23. $W = \{\mathbf{x} : \mathbf{x} = \begin{bmatrix} a \\ 0 \end{bmatrix}, a \text{ any real number }\}$.

25. $W = \{\mathbf{x} : \mathbf{x} = \begin{bmatrix} a \\ 2 \end{bmatrix}, a \text{ any real number }\}$.

27. $W = \{\mathbf{x} : \mathbf{x} = \begin{bmatrix} x_1 \\ x_2 \\ x_3 \end{bmatrix}, x_1 + x_2 - 2x_3 = 0\}$.

29. $W = \{\mathbf{x} : \mathbf{x} = \begin{bmatrix} 0 \\ x_2 \\ x_3 \end{bmatrix}, x_2, x_3 \text{ any real number }\}$.

2.2 Vector Space Properties of R^n

1. Clearly θ is in W . Suppose $\mathbf{u}$ and $\mathbf{v}$ are in W , where $\mathbf{u} = \begin{bmatrix} u_1 \\ u_2 \end{bmatrix}$ and $\mathbf{v} = \begin{bmatrix} v_1 \\ v_2 \end{bmatrix}$. Then $u_1 = 2u_2$ and $v_1 = 2v_2$. If a is any scalar then $\mathbf{u}+\mathbf{v} = \begin{bmatrix} u_1 + v_1 \\ u_2 + v_2 \end{bmatrix}$ and $a\mathbf{u} = \begin{bmatrix} au_1 \\ au_2 \end{bmatrix}$. But $u_1 + v_1 = 2u_2 + 2v_2 = 2(u_2 + v_2)$ and $au_1 = a(2u_2) = 2(au_2)$, so $\mathbf{u}+\mathbf{v}$ and $a\mathbf{u}$ are in W. By Theorem 2, W is a subspace of R^2. Geometrically, W consists of the points on the line with equation $x = 2y$.

3. W is not a subspace of R^2 since, for example, $\mathbf{u} = \begin{bmatrix} 1 \\ 1 \end{bmatrix}$ and $\mathbf{v} = \begin{bmatrix} 1 \\ -1 \end{bmatrix}$ are in W whereas $\mathbf{u}+\mathbf{v}$ is not in W. Note that W satisfies properties (s1) and (s3) of Theorem 2.

5. Clearly, $\theta = \begin{bmatrix} 0 \\ 0 \end{bmatrix}$ is in W. If $\mathbf{u} = \begin{bmatrix} u_1 \\ u_2 \end{bmatrix}$ and $\mathbf{v} = \begin{bmatrix} v_1 \\ v_2 \end{bmatrix}$ are in W, then $u_1 = v_1 = 0$. Thus $\mathbf{u}+\mathbf{v} = \begin{bmatrix} 0 \\ u_2 + v_2 \end{bmatrix}$ is in W. Likewise, if a is any scalar, $a\mathbf{u} = \begin{bmatrix} 0 \\ au_2 \end{bmatrix}$ is in W. By Theorem 2, W is a subspace of R^2. Geometrically, W consists of the points on the y - axis .

7. W satisfies none of the properties (s1) - (s3) of Theorem 2, so W is not a subspace of R^2. Clearly, $\theta = \begin{bmatrix} 0 \\ 0 \end{bmatrix}$ is not in W. Also, $\mathbf{u} = \begin{bmatrix} 1 \\ 0 \end{bmatrix}$ and $\mathbf{v} = \begin{bmatrix} 0 \\ 1 \end{bmatrix}$ are in W whereas $\mathbf{u}+\mathbf{v}$ is not in W. Finally,

   if $a \neq 1$ then $a\mathbf{v}$ is not in W.

9. Clearly $\theta = \begin{bmatrix} 0 \\ 0 \\ 0 \end{bmatrix}$ is in W. If $\mathbf{u} = \begin{bmatrix} u_1 \\ u_2 \\ u_3 \end{bmatrix}$ and $\mathbf{v} = \begin{bmatrix} v_1 \\ v_2 \\ v_3 \end{bmatrix}$ are in W and a is any scalar, then $u_3 = 2u_1 - u_2$ and $v_3 = 2v_1 - v_2$. Now $u_3 + v_3 = (2u_1 - u_2) + (2v_1 - v_2) = 2(u_1 + v_1) - (u_2 + v_2)$ and $au_3 = a(2u_1 - u_2) = 2(au_1) - au_2$. Therefore $\mathbf{u}+\mathbf{v}$ and $a\mathbf{u}$ are in W. By Theorem 2, W is a subspace of R^3. Geometrically, W consists of the points on the plane with equation $2x - y - z = 0$.

11. W is not a subspace of R^3. For example, $\mathbf{u} = \begin{bmatrix} 1 \\ 1 \\ 1 \end{bmatrix}$ and $\mathbf{v} = \begin{bmatrix} 1 \\ 0 \\ 0 \end{bmatrix}$ are in W but $\mathbf{u}+\mathbf{v}$ is not in W. Also if $a \neq 1$ and $a \neq 0$ then $a\mathbf{u}$ is not in W.

13. W is not a subspace of R^3. For example $\mathbf{u} = \begin{bmatrix} 1 \\ 0 \\ 0 \end{bmatrix}$ and $\mathbf{v} = \begin{bmatrix} 2 \\ 2 \\ 0 \end{bmatrix}$ are in W but $\mathbf{u}+\mathbf{v}$ is not in W. Also, if $a \neq 1$ and $a \neq 0$ then $a\mathbf{u}$ is not in W.

15. Clearly θ is in W. If $\mathbf{u}$ and $\mathbf{v}$ are vectors in W then $\mathbf{u}$ and $\mathbf{v}$ can be expressed in the form $\mathbf{u} = \begin{bmatrix} 2a \\ -a \\ a \end{bmatrix}$ and $\mathbf{v} = \begin{bmatrix} 2b \\ -b \\ b \end{bmatrix}$. Then $\mathbf{u}+\mathbf{v} = \begin{bmatrix} 2(a+b) \\ -(a+b) \\ a+b \end{bmatrix}$. Similarly, for any scalar c, $c\mathbf{u} = \begin{bmatrix} 2ca \\ -ca \\ ca \end{bmatrix}$. By Theorem 2, W is a subspace of R^3. Geometrically W consists of the points on the line with parametric equations $x = 2t,\ y = -t,\ z = t$.

17. Clearly θ is in W. Moreover, any two elements $\mathbf{u}$ and $\mathbf{v}$ in W can be written in the form $\mathbf{u} = \begin{bmatrix} a \\ 0 \\ 0 \end{bmatrix}$ and $\mathbf{v} = \begin{bmatrix} b \\ 0 \\ 0 \end{bmatrix}$. Therefore $\mathbf{u}+\mathbf{v} = \begin{bmatrix} a+b \\ 0 \\ 0 \end{bmatrix}$ is in W and for any scalar c, $c\mathbf{u} = \begin{bmatrix} ca \\ 0 \\ 0 \end{bmatrix}$ is in W. By Theorem 2, W is a subspace of R^3. Geometrically, W consists of the points on the $x-$ axis.

19. The vector $\mathbf{u} = \begin{bmatrix} u_1 \\ u_2 \\ u_3 \end{bmatrix}$ is in W if and only if $0 = \mathbf{a}^T\mathbf{u} = u_1 + 2u_2 + u_3$. Thus geometrically W consists of the points in R^3 which lie in the plane $x + 2y + 3z = 0$.

21. Clearly, $\mathbf{a}^T\theta = \mathbf{b}^T\theta = 0$, so θ is in W. Let $\mathbf{a}^T\mathbf{u} = \mathbf{b}^T\mathbf{u} = \mathbf{a}^T\mathbf{v} = \mathbf{b}^T\mathbf{v} = 0$. It then follows that $\mathbf{a}^T(\mathbf{u}+\mathbf{v}) = 0$ and $\mathbf{b}^T(\mathbf{u}+\mathbf{v}) = 0$. Therefore $\mathbf{u}+\mathbf{v}$ is in W. Likewise, $\mathbf{a}^T(c\mathbf{u}) = c(\mathbf{a}^T\mathbf{u}) = 0$ and $\mathbf{b}^T(c\mathbf{u}) =$

$c(\mathbf{b}^T\mathbf{u}) = 0$ for any scalar c. Therefore $c\mathbf{u}$ is in W. Thus W is a subspace of R^3.

23. The vector $\mathbf{u} = \begin{bmatrix} u_1 \\ u_2 \\ u_3 \end{bmatrix}$ is in W if and only if $0 = \mathbf{a}^T\mathbf{u} = u_1 + 2u_2 + 2u_3$ and $0 = \mathbf{b}^T\mathbf{u} = u_1 + 3u_2$. Thus W is the set of points on the line formed by the intersecting planes $x + 2y + 2z = 0$ and $x + 3y = 0$. Solving yields $x = -6z$ and $y = 2z$ so the line has parametric equations $x = -6t,\ y = 2t,\ z = t$.

25. The vector $\mathbf{u} = \begin{bmatrix} u_1 \\ u_2 \\ u_3 \end{bmatrix}$ is in W if and only if $0 = \mathbf{a}^T\mathbf{u} = u_1 - u_3$ and $0 = \mathbf{b}^T\mathbf{u}$ $= -2u_1 + 2u_3$. Clearly the latter condition is redundant so W consists of the points in the plane $x - z = 0$.

27. Property (m1) is not satisfied. For example $3(2\mathbf{x}) = 3\begin{bmatrix} 4x_1 \\ 4x_2 \end{bmatrix} = \begin{bmatrix} 24x_1 \\ 24x_2 \end{bmatrix}$ where $6\mathbf{x} = \begin{bmatrix} 12x_1 \\ 12x_2 \end{bmatrix}$. Also, (m4) is not satisfied since $1\mathbf{x} = \begin{bmatrix} 2x_1 \\ 2x_2 \end{bmatrix} \neq \mathbf{x}$.

29. The set of points on the line can be expressed as the set $W = \{t\begin{bmatrix} a \\ b \\ c \end{bmatrix} : t$ any real number $\}$. Taking $t = 0$ we see that $\boldsymbol{\theta}$ is in W. If $\mathbf{u} = r\begin{bmatrix} a \\ b \\ c \end{bmatrix}$ and $\mathbf{v} = s\begin{bmatrix} a \\ b \\ c \end{bmatrix}$ then $\mathbf{u} + \mathbf{v} = (r + s)\begin{bmatrix} a \\ b \\ c \end{bmatrix}$ is in W. Likewise, if k is any scalar then $k\mathbf{u} = kr\begin{bmatrix} a \\ b \\ c \end{bmatrix}$ is in W. Therefore W is a subspace of R^3.

2.3 Examples of Subspaces

1. By definition $\text{Sp}(S) = \{t\begin{bmatrix} 1 \\ -1 \end{bmatrix} : t$ any real number $\}$. Thus if $\mathbf{x} = \begin{bmatrix} x_1 \\ x_2 \end{bmatrix}$ is in R^2 , then $\mathbf{x}$ is in $\text{Sp}(S)$ if and only if $x_1 + x_2 = 0$. In particular $\text{Sp}(S)$ is the line with equation $x + y = 0$.

3. $\text{Sp}(S) = \{t\begin{bmatrix} 0 \\ 0 \end{bmatrix} : t$ any real number $\} = \{\mathbf{e}\}$.$\text{Sp}(S)$ is the point $(0, 0)$.

5. $\text{Sp}(S) = \{\mathbf{x}$ in $R^2 : \mathbf{x} = k_1\mathbf{a} + k_2\mathbf{d}$ for scalars k_1 and $k_2\}$. For an arbitrary vector $\mathbf{x}$ in R^2, $\mathbf{x} = \begin{bmatrix} x_1 \\ x_2 \end{bmatrix}$, the equation $k_1\mathbf{a} + k_2\mathbf{d} = \mathbf{x}$ has augmented matrix $\begin{bmatrix} 1 & 1 & x_1 \\ -1 & 0 & x_2 \end{bmatrix}$.
This matrix reduces to
$\begin{bmatrix} 1 & 1 & x_1 \\ 0 & 1 & x_1 + x_2 \end{bmatrix}$ and backsolving yields $k_1 = -x_2$, and $k_2 = x_1 + x_2$. It follows that $\text{Sp}(S) = R^2$.

7. $\text{Sp}(S) = \{\mathbf{x}$ in $R^2 : \mathbf{x} = k_1\mathbf{b} + k_2\mathbf{e}$ for scalars k_1 and $k_2\}$. But $k_1\mathbf{b} + k_2\mathbf{e} = k_1\mathbf{b}$ so $\text{Sp}(S) = \text{Sp}(\{\mathbf{b}\})$. It follows that $\text{Sp}(S) = \{ t \begin{bmatrix} 2 \\ -3 \end{bmatrix} : t$ any real number $\}$. If $\mathbf{x} = \begin{bmatrix} x_1 \\ x_2 \end{bmatrix}$ is in R^2, $\mathbf{x}$ is in $\text{Sp}(S)$ if and only if $3x_1 + 2x_2 = 0$. Thus, $\text{Sp}(S) = \{\mathbf{x} : 3x_1 + 2x_2 = 0\}$; so $\text{Sp}(S)$ is the line with equation $3x + 2y = 0$.

9. $\text{Sp}(S) = \{\mathbf{x}$ in $R^2 : \mathbf{x} = k_1\mathbf{b} + k_2\mathbf{c} + k_3\mathbf{d}$ for scalars $k_1, k_2, k_3\}$. With $\mathbf{x} = \begin{bmatrix} x_1 \\ x_2 \end{bmatrix}$, the equation $k_1\mathbf{b} + k_2\mathbf{c} + k_3\mathbf{d} = \mathbf{x}$ has augmented matrix $\begin{bmatrix} 2 & -2 & 1 & x_1 \\ -3 & 2 & 0 & x_2 \end{bmatrix}$. The reduction reveals that the system is consistent for every $\mathbf{x}$, so $\text{Sp}(S) = R^2$.

11. Since $k_1\mathbf{a} + k_2\mathbf{c} + k_3\mathbf{e} = k_1\mathbf{a} + k_2\mathbf{c}$, $\text{Sp}(S) = \text{Sp}\{\mathbf{a}, \mathbf{c}\}$. But $\text{Sp}\{\mathbf{a}, \mathbf{c}\} = \{\mathbf{x} : x_1 + x_2 = 0\}$. Thus $\text{Sp}(S)$ is the line with equation $x + y = 0$. (cf. Exercise 6).

13. $\text{Sp}(S) = \{\mathbf{x}$ in $R^3 : \mathbf{x} = t\mathbf{w}$ for some scalar $t\}$. Therefore $\text{Sp}(S)$ is the line through $(0,0,0)$ and $(0,-1,1)$. The parametric equations are $x = 0$, $y = -t$, $z = t$. If $\mathbf{x} = [x_1, x_2, x_3]^T$ then $t\mathbf{w} = \mathbf{x}$ has augmented matrix $\begin{bmatrix} 0 & x_1 \\ -1 & x_2 \\ 1 & x_3 \end{bmatrix}$ which reduces to $\begin{bmatrix} 1 & x_3 \\ 0 & x_2 + x_3 \\ 0 & x_1 \end{bmatrix}$.

The system is consistent if $x_2 + x_3 = 0$ and $x_1 = 0$ so we have $\text{Sp}(S) = \{\mathbf{x} : x_2 + x_3 = 0$ and $x_1 = 0\}$. Therefore $\text{Sp}(S)$ is the line formed by the the intersecting planes $y + z = 0$ and $x = 0$.

15. $\text{Sp}(S) = \{\mathbf{u}$ in $R^3 : \mathbf{u} = k_1\mathbf{v} + k_2\mathbf{x}\}$. The equation $k_1\mathbf{v} + k_2\mathbf{x} = \mathbf{u}$ has augmented matrix

$$\begin{bmatrix} 1 & 1 & u_1 \\ 2 & 1 & u_2 \\ 0 & -1 & u_3 \end{bmatrix}$$

, which reduces to

$$\begin{bmatrix} 1 & 1 & u_1 \\ 0 & 1 & 2u_1 - u_2 \\ 0 & 0 & 2u_1 - u_2 + u_3 \end{bmatrix}$$

. A solution exists if and only if $2u_1 - u_2 + u_3 = 0$, so $\text{Sp}(S) = \{\mathbf{u} : 2u_1 - u_2 + u_3 = 0\}$. $\text{Sp}(S)$ is the plane with equation $2x - y + z = 0$.

17. $\text{Sp}(S) = \{\mathbf{u}$ in $R^3 : \mathbf{u} = k_1\mathbf{w} + k_2\mathbf{x} + k_3\mathbf{z}\}$. For $\mathbf{u} = [u_1, u_2, u_3]^T$ the equation $k_1\mathbf{w} + k_2\mathbf{x} + k_3\mathbf{z} = \mathbf{u}$ has augmented matrix

$$\begin{bmatrix} 0 & 1 & 1 & u_1 \\ -1 & 1 & 0 & u_2 \\ 1 & -1 & 2 & u_3 \end{bmatrix}$$

. This matrix reduces to

$$\begin{bmatrix} 1 & -1 & 2 & u_3 \\ 0 & 1 & 1 & u_1 \\ 0 & 0 & 2 & u_2 + u_3 \end{bmatrix}$$

. Since the system is consistent for every $\mathbf{u}$ in R^3, $\text{Sp}(S) = R^3$.

19. The matrix $\begin{bmatrix} 0 & 1 & -2 & u_1 \\ -1 & 1 & -2 & u_2 \\ 1 & -1 & 2 & u_3 \end{bmatrix}$ reduces to

$\begin{bmatrix} 1 & -1 & 2 & u_3 \\ 0 & 1 & -2 & u_1 \\ 0 & 0 & 0 & u_2 + u_3 \end{bmatrix}$ so the system of equations $k_1\mathbf{w} + k_2\mathbf{x} + k_3\mathbf{y} = \mathbf{u}$ is consistent if and only if $u_2 + u_3 = 0$. Therefore $\mathrm{Sp}(S) = \{\mathbf{u} : u_2 + u_3 = 0\}$ and $\mathrm{Sp}(S)$ is the plane with equation $y + z = 0$.

21. By Exercise 15, $\mathrm{Sp}(S) = \{\mathbf{u} \text{ in } R^3 : 2u_1 - u_2 + u_3 = 0\}$. Thus the vectors given in (b), (c), and (e) are in $\mathrm{Sp}(S)$. From the calculations done in Exercise 15, it follows that when the system of equations $k_1\mathbf{v} + k_2\mathbf{x} = \mathbf{u}$ is consistent, the unique solution is $k_1 = -u_1 + u_2$ and $k_2 = 2u_1 - u_2$. Thus in (b), $[1, 1, -1]^T = \mathbf{x}$; in (c), $[1, 2, 0]^T = \mathbf{v}$; and in (e), $[-1, 2, 4]^T = 3\mathbf{v} - 4\mathbf{x}$.

23. The vectors $\mathbf{d}$ and $\mathbf{e}$ are in $\mathcal{N}(A)$ since by direct calculation, $A\mathbf{d} = \theta$ and $A\mathbf{e} = \theta$.

25. The vectors $\mathbf{x}$ and $\mathbf{y}$ are in $\mathcal{N}(A)$ since, by direct calculation $A\mathbf{x} = \theta$ and $A\mathbf{y} = \theta$.

27. The matrix $[A \mid \mathbf{b}]$ is row equivalent to the matrix :

$$\begin{bmatrix} -1 & 3 & b_1 \\ 0 & 0 & 2b_1 + b_2 \end{bmatrix}.$$

It follows that the homogeneous system $A\mathbf{x} = \theta$ has solution $x_1 = 3x_2$ whereas the system $A\mathbf{x} = \mathbf{b}$ is consistent if and only if $2b_1 + b_2 = 0$. Therefore $\mathcal{N}(A) = \{\mathbf{x} : x_1 - 3x_2 = 0\}$ and $\mathcal{R}(A) =$

$\{\mathbf{b} : 2b_1 + b_2 = 0\}$.

29. The matrix $[A \mid \mathbf{b}]$ reduces to $\begin{bmatrix} 1 & 1 & b_1 \\ 0 & 3 & -2b_1 + b_3 \end{bmatrix}$. Setting $\mathbf{b} = \theta$ and solving yields $x_1 = x_2 = 0$, so $\mathcal{N}(A) = \{\theta\}$. Since the system $A\mathbf{x} = \mathbf{b}$ is consistent for every $\mathbf{b}$ in $R^2, \mathcal{R}(A) = R^2$.

31. The matrix $[A \mid \mathbf{b}]$ reduces to $\begin{bmatrix} 1 & 2 & 1 & b_1 \\ 0 & 0 & 1 & -3b_1 + b_2 \end{bmatrix}$. Setting $\mathbf{b} = \theta$ and backsolving yields $x_1 = -2x_2$, $x_3 = 0$ as the solution to $A\mathbf{x} = \theta$. Thus $\mathcal{N}(A) = \{\mathbf{x} \text{ in } R^3 : x_1 + 2x_2 = 0$ and $x_3 = 0\}$. Since $A\mathbf{x} = \mathbf{b}$ is consistent for arbitrary $\mathbf{b}$ in R^2, $\mathcal{R}(A) = R^2$.

33. The matrix $[A \mid \mathbf{b}]$ reduces to $\begin{bmatrix} 0 & 1 & b_1 \\ 0 & 0 & -2b_1 + b_2 \\ 0 & 0 & -3b_1 + b_3 \end{bmatrix}$. Setting $\mathbf{b} = \theta$

yields $x_2 = 0$, x_1 arbitrary as the solution to $A\mathbf{x} = \theta$. Thus $\mathcal{N}(A) = \{\mathbf{x}$ in $R^2 : x_2 = 0\}$. The system $A\mathbf{x} = \mathbf{b}$ is consistent if and only if $-2b_1 + b_2 = 0$ and $-3b_1 + b_3 = 0$. Therefore $\mathcal{R}(A) =$

$\{\mathbf{b}$ in $R^3 : b_2 = 2b_1$ and $b_3 = 3b_1\}$.

35. The matrix $[A \mid \mathbf{b}]$ reduces to $\begin{bmatrix} 1 & 2 & 3 & b_1 \\ 0 & 1 & -2 & -b_1 + b_2 \\ 0 & 0 & 0 & -4b_1 + 2b_2 + b_3 \end{bmatrix}$.

Setting $\mathbf{b} = \theta$ and backsolving the reduced system yields $\mathcal{N}(A) = \{\mathbf{x}$ in $R^3 : x_1 = -7x_3$ and $x_2 = 2x_3\}$. The system $A\mathbf{x} = \mathbf{b}$ is consistent if and only if $-4b_1 + 2b_2 + b_3 = 0$ so $\mathcal{R}(A) = \{\mathbf{b}$ in $R^3 : -4b_1 + 2b_2 + b_3 = 0\}$.

37. The matrix $[A \mid \mathbf{b}]$ reduces to $\begin{bmatrix} 1 & 2 & 1 & b_1 \\ 0 & 1 & 2 & -2b_1 + b_2 \\ 0 & 0 & 1 & b_1 - b_2 + b_3 \end{bmatrix}$. Setting $\mathbf{b} = \theta$ and solving yields $\mathcal{N}(A) = \{\theta\}$. The system $A\mathbf{x} = \mathbf{b}$ is consistent for all $\mathbf{b}$ so $\mathcal{R}(A) = R^3$.

39. (a) From the description of $\mathcal{R}(A)$ obtained in Exercise 27 it follows that the vectors $\mathbf{b}$ in (ii),(v), and (vi) are in $\mathcal{R}(A)$.

(b) When the system of equations $A\mathbf{x} = \mathbf{b}$ is consistent, the calculations done in Exercise 27 show that the solution is given by $x_1 = -b_1 + 3x_2$, where x_2 is arbitrary. Thus for (ii), $\mathbf{x} = [1, 0]^T$ is one choice; for (v), $\mathbf{x} = [0, 1]^T$ is one choice; for (vi), $\mathbf{x} = [0, 0]^T$ is one choice.

(c) If $A\mathbf{x} = \mathbf{b}$, where $\mathbf{x} = [x_1, x_2]^T$, then $\mathbf{b} = x_1\mathbf{A}_1 + x_2\mathbf{A}_2$. Therefore for (ii), $\mathbf{b} = \mathbf{A}_1$; for (v), $\mathbf{b} = \mathbf{A}_2$; for (vi), $\mathbf{b} = 0\mathbf{A}_1 + 0\mathbf{A}_2$.

41. (a) From the description of $\mathcal{R}(A)$ obtained in Exercise 35, the vectors $\mathbf{b}$ in (i), (iii), (v), and (vi), are in $\mathcal{R}(A)$.

(b) When the system $A\mathbf{x} = \mathbf{b}$ is consistent, the solution is given by $x_1 = 3b_1 - 2b_2 - 7x_3$ and $x_2 = -b_1 + b_2 + 2x_3$, where x_3 is arbitrary. Thus for (i), $\mathbf{x} = [-1, 1, 0]^T$ is one choice; for (iii), $\mathbf{x} = [-2, 3, 0]^T$ is one choice for (v), $\mathbf{x} = [-2, 1, 0]^T$ is one choice; for (vi), $\mathbf{x} = [0, 0, 0]^T$ is one choice.

(c). If $A\mathbf{x} = \mathbf{b}$, where x= $[x_1, x_2, x_3]^T$, then $\mathbf{b} = x_1\mathbf{A}_1 + x_2\mathbf{A}_2 + x_3\mathbf{A}_3$.
Thus it follows from (b) that for (i), $\mathbf{b} = -\mathbf{A}_1 + \mathbf{A}_2$; for (iii), $\mathbf{b} = -2\mathbf{A}_1 + 3\mathbf{A}_2$; for (v), $\mathbf{b} = -2\mathbf{A}_1 + \mathbf{A}_2$; for (vi), $\mathbf{b} = 0\mathbf{A}_1 + 0\mathbf{A}_2 + 0\mathbf{A}_3$,

43. $A = [\ 3 \quad -4 \quad 2\ ]$

45. $A = [\mathbf{w}, \mathbf{x}, \mathbf{z}] = \begin{bmatrix} 0 & 1 & 1 \\ -1 & 1 & 0 \\ 1 & -1 & 2 \end{bmatrix}$

47. Let A be the (3 x 3) matrix whose columns are the vectors given in S. Then A^T reduces to $B^T = \begin{bmatrix} -2 & 1 & 3 \\ 0 & 3 & 2 \\ 0 & 0 & 0 \end{bmatrix}$. The nonzero columns of B, $\mathbf{w_1} = [-2,1,3]^T$ and $\mathbf{w_2} = [0,3,2]^T$ form a basis for Sp(S).

49. Let A be the (3 x 4) matrix whose columns are the vectors given in S. Then A^T reduces to $B^T = \begin{bmatrix} 1 & 2 & 2 \\ 0 & 3 & 1 \\ 0 & 0 & 0 \\ 0 & 0 & 0 \end{bmatrix}$. The nonzero columns of B, $\mathbf{w_1} = [1,2,2]^T$ and $\mathbf{w_2} = [0,3,1]^T$ form a basis for Sp(S).

2.4 Bases for Subspaces

1. Backsolving the given system yields $x_1 = x_3 - x_4$, and $x_2 = x_4$. Thus

$$\begin{bmatrix} x_1 \\ x_2 \\ x_3 \\ x_4 \end{bmatrix} = \begin{bmatrix} x_3 - x_4 \\ x_4 \\ x_3 \\ x_4 \end{bmatrix} = x_3 \begin{bmatrix} 1 \\ 0 \\ 1 \\ 0 \end{bmatrix} + x_4 \begin{bmatrix} -1 \\ 1 \\ 0 \\ 1 \end{bmatrix}.$$

As in Example 5,$\{[1,0,1,0]^T, [-1,1,0,1]^T\}$ is a basis for W.

3. Writing $x_1 = x_2 - x_3 + 3x_4$ we have

$$\begin{bmatrix} x_1 \\ x_2 \\ x_3 \\ x_4 \end{bmatrix} = \begin{bmatrix} x_2 - x_3 + 3x_4 \\ x_2 \\ x_3 \\ x_4 \end{bmatrix} = x_2 \begin{bmatrix} 1 \\ 1 \\ 0 \\ 0 \end{bmatrix} + x_3 \begin{bmatrix} -1 \\ 0 \\ 1 \\ 0 \end{bmatrix} + x_4 \begin{bmatrix} 3 \\ 0 \\ 0 \\ 1 \end{bmatrix}.$$

Thus $\{[1,1,0,0]^T, [-1,0,1,0]^T, [3,0,0,1]^T\}$ is the desired basis.

5. Since $x_1 = -x_2$ we have

$$\begin{bmatrix} x_1 \\ x_2 \\ x_3 \\ x_4 \end{bmatrix} = \begin{bmatrix} -x_2 \\ x_2 \\ x_3 \\ x_4 \end{bmatrix} = x_2 \begin{bmatrix} -1 \\ 1 \\ 0 \\ 0 \end{bmatrix} + x_3 \begin{bmatrix} 0 \\ 0 \\ 1 \\ 0 \end{bmatrix} + x_4 \begin{bmatrix} 0 \\ 0 \\ 0 \\ 1 \end{bmatrix}.$$

It follows that $\{[-1,1,0,0]^T, [0,0,1,0]^T, [0,0,0,1]^T\}$ is a basis for W.

7. Backsolving yields $x_1 = -2x_3 - x_4$ and $x_2 = -x_3$. Thus

$$\begin{bmatrix} x_1 \\ x_2 \\ x_3 \\ x_4 \end{bmatrix} = \begin{bmatrix} -2x_3 - x_4 \\ -x_3 \\ x_3 \\ x_4 \end{bmatrix} = x_3 \begin{bmatrix} -2 \\ -1 \\ 1 \\ 0 \end{bmatrix} + x_4 \begin{bmatrix} -1 \\ 0 \\ 0 \\ 1 \end{bmatrix}.$$

Therefore $\{[-2,-1,1,0]^{\mathrm{T}}, [-1,0,0,1]^{\mathrm{T}}\}$ is a basis for W.

9. Let $\{\mathbf{w}_1, \mathbf{w}_2\}$ be the basis found in Exercise 1. (a) $\mathbf{x} = 2\mathbf{w}_1 + \mathbf{w}_2$ (b) $\mathbf{x}$ is not in W. (c) $\mathbf{x} = -3\mathbf{w}_2$ (d) $\mathbf{x} = 2\mathbf{w}_1$.

11. (a) $B = \begin{bmatrix} 1 & 2 & 3 & -1 \\ 0 & -1 & -1 & 1 \\ 0 & 0 & 0 & 0 \end{bmatrix}$.

(b) Backsolving the reduced system $B\mathbf{x} = \theta$ yields the solution $x_1 = -x_3 - x_4, x_2 = -x_3 + x_4$ for the homogeneous system $A\mathbf{x} = \theta$. Thus $\mathbf{x} = [x_1, x_2, x_3, x_4]^{\mathrm{T}}$ is in $\mathcal{N}(A)$ if and only if

$$\begin{bmatrix} x_1 \\ x_2 \\ x_3 \\ x_4 \end{bmatrix} = \begin{bmatrix} -x_3 - x_4 \\ -x_3 + x_4 \\ x_3 \\ x_4 \end{bmatrix} = x_3 \begin{bmatrix} -1 \\ -1 \\ 1 \\ 0 \end{bmatrix} + x_4 \begin{bmatrix} -1 \\ 1 \\ 0 \\ 1 \end{bmatrix}$$

It follows that $\{[-1,-1,1,0]^T, [-1,1,0,1]^T\}$ is a basis for $\mathcal{N}(A)$.

(c) It follows from (b) that $x_1\mathbf{A_1} + x_2\mathbf{A_2} + x_3\mathbf{A_3} + x_4\mathbf{A_4} = \theta$ if and only if $x_1 = -x_3 - x_4$ and $x_2 = -x_3 + x_4$. Since x_3 and x_4 are unconstrained variables $\{\mathbf{A}_1, \mathbf{A}_2\}$ is a basis for $\mathcal{R}(A)$. Setting $x_3 = 1$ and $x_4 = 0$ yields $x_1 = -1$ and $x_2 = -1$ so $-\mathbf{A_1} - \mathbf{A_2} + \mathbf{A_3} = \theta$. Therefore $\mathbf{A_3} = \mathbf{A_1} + \mathbf{A_2}$. Similarly, setting $x_3 = 0$ and $x_4 = 1$ yields $\mathbf{A_4} = \mathbf{A_1} - \mathbf{A_2}$.

(d) The nonzero rows of B form a basis for the row space of A ; that is $\{[1,2,3,-1], [0,-1,-1,1]\}$ is the desired basis.

13. (a) $B = \begin{bmatrix} 1 & 2 & 1 & 0 \\ 0 & 1 & 1 & -1 \\ 0 & 0 & 0 & 0 \\ 0 & 0 & 0 & 0 \end{bmatrix}$.

(b) The homogeneous system $A\mathbf{x} = \theta$ has solution $x_1 = x_3 - 2x_4, x_2 = -x_3 + x_4$. Thus $\mathbf{x} = [x_1, x_2, x_3, x_4]^{\mathrm{T}}$ is in $\mathcal{N}(A)$ if and only if

$$\begin{bmatrix} x_1 \\ x_2 \\ x_3 \\ x_4 \end{bmatrix} = \begin{bmatrix} x_3 - 2x_4 \\ -x_3 + x_4 \\ x_3 \\ x_4 \end{bmatrix} = x_3 \begin{bmatrix} 1 \\ -1 \\ 1 \\ 0 \end{bmatrix} + x_4 \begin{bmatrix} -2 \\ 1 \\ 0 \\ 1 \end{bmatrix}.$$

The set $\{[1,-1,1,0]^{\mathrm{T}}, [-2,1,0,1]^{\mathrm{T}}\}$ is a basis for $\mathcal{N}(A)$.

(c) It follows from (b) that in the equation $x_1\mathbf{A_1} + x_2\mathbf{A_2} + x_3\mathbf{A_3} + x_4\mathbf{A_4} = \theta, x_3$ and x_4 are unconstrained variables. Therefore $\{\mathbf{A_1}, \mathbf{A_2}\}$ is a basis for $\mathcal{R}(A)$. Furthermore $\mathbf{A_1} - \mathbf{A_2} + \mathbf{A_3} = \theta$, so $\mathbf{A_3} = -\mathbf{A_1} + \mathbf{A_2}$. Likewise $-2\mathbf{A_1} + \mathbf{A_2} + \mathbf{A_4} = \theta$, so $\mathbf{A_4} = 2\mathbf{A_1} - \mathbf{A_2}$.

(d) The nonzero rows of B, $[1,2,1,0], [0,1,1,-1]$, form a basis for the row space of A.

15. (a) $B = \begin{bmatrix} 1 & 2 & 1 \\ 0 & 0 & -1 \\ 0 & 0 & 0 \end{bmatrix}.$

(b) The system $A\mathbf{x} = \theta$ has solution $x_1 = -2x_2, x_3 = 0$. Thus $\mathcal{N}(A) = \{\mathbf{x} : \mathbf{x} = [-2x_2, x_2, 0]^{\mathrm{T}}\}$ and $\{[-2,1,0]^{\mathrm{T}}\}$ is a basis for $\mathcal{N}(A)$.

(c) In the equation $x_1\mathbf{A_1} + x_2\mathbf{A_2} + x_3\mathbf{A_3} = \theta$, x_2 is an unconstrained variable, so $\{\mathbf{A_1}, \mathbf{A_3}\}$ is a basis for $\mathcal{R}(A)$.

Furthermore, $-2\mathbf{A_1} + \mathbf{A_2} = \theta$, so $\mathbf{A_2} = 2\mathbf{A_1}$.

(d) $\{[1,2,1], [0,0,-1]\}$ is a basis for the row space of A.

17. The matrix A^{T} is row equivalent to $B^{\mathrm{T}} = \begin{bmatrix} 1 & 3 & 1 \\ 0 & -1 & -1 \\ 0 & 0 & 0 \\ 0 & 0 & 0 \end{bmatrix}$. The desired basis is $\{[1,3,1]^{\mathrm{T}}, [0,-1,-1]^{\mathrm{T}}\}$, formed by taking the nonzero columns of B.

19. The matrix A^{T} is row equivalent to $B^{\mathrm{T}} = \begin{bmatrix} 1 & 2 & 2 & 0 \\ 0 & 1 & -2 & 1 \\ 0 & 0 & 0 & 0 \\ 0 & 0 & 0 & 0 \end{bmatrix}$ so $\{[1,2,2,0]^{\mathrm{T}}, [0,1,-2,1]^{\mathrm{T}}\}$ is a basis for $\mathcal{R}(A)$.

21. (a) For the given vectors $\mathbf{u_1}$ and $\mathbf{u_2}$ the equation $x_1\mathbf{u_1} + x_2\mathbf{u_2} = \theta$ has solution $x_1 = -2x_2$ where x_2 is an unconstrained variable. Therefore $\{\mathbf{u_1}\}$ is a basis for Sp(S), where $\mathbf{u_1} = [1,2]^{\mathrm{T}}$.

(b) If $A = [\mathbf{u}_1, \mathbf{u}_2]$ then A^T is row equivalent to $B^T = \begin{bmatrix} 1 & 2 \\ 0 & 0 \end{bmatrix}$. Therefore $\{[1,2]^T\}$ is a basis for Sp(S).

23. (a) For the given vectors $\mathbf{u_1}, \mathbf{u_2}, \mathbf{u_3}, \mathbf{u_4}$ the equation $x_1\mathbf{u_1} + x_2\mathbf{u_2} + x_3\mathbf{u_3} + x_4\mathbf{u_4} = \theta$ has solution $x_1 = -x_3 - 3x_4, x_2 = -x_3 + x_4$. Since x_3 and x_4 are unconstrained variables, $\{\mathbf{u_1}, \mathbf{u_2}\}$ is a basis for Sp(S), where $\mathbf{u_1} = [1,2,1]^{\mathrm{T}}$ and $\mathbf{u_2} = [2,5,0]^{\mathrm{T}}$.

(b) If $A = [\mathbf{u_1}, \mathbf{u_2}, \mathbf{u_3}, \mathbf{u_4}]$ then A^{T} is row equivalent to $B^{\mathrm{T}} = \begin{bmatrix} 1 & 2 & 1 \\ 0 & 1 & -2 \\ 0 & 0 & 0 \\ 0 & 0 & 0 \end{bmatrix}$.

Therefore $\{[1,2,1]^{\mathrm{T}}, [0,1,-2]^{\mathrm{T}}\}$ is a basis for $\mathrm{Sp}(S)$.

25. (a) Let A denote the given matrix. The homogeneous system $A\mathbf{x} = \theta$ has solution $x_1 = 0, x_2$ is arbitrary, $x_3 = 0$. Thus $\{[0,1,0]^{\mathrm{T}}\}$ is a basis for $\mathcal{N}(A)$.

(b) Let A denote the given matrix. The system $A\mathbf{x} = \theta$ has solution $x_1 = -x_2$, where x_2 and x_3 are arbitrary. Thus $\{[-1,1,0]^{\mathrm{T}}, [0,0,1]^{\mathrm{T}}\}$ is a basis for $\mathcal{N}(A)$.

(c) The system $A\mathbf{x} = \theta$ has solution $x_1 = -x_2, x_3 = 0$, where x_2 is arbitrary. The set $\{[-1,1,0]^{\mathrm{T}}\}$ is a basis for $\mathcal{N}(A)$.

27. The equation $x_1\mathbf{v_1} + x_2\mathbf{v_2} + x_3\mathbf{v_3} = \theta$ has solution $x_1 = -2x_3$, $x_2 = -3x_3, x_3$ arbitrary. In particular, $x_1 = -2, x_2 = -3, x_3 = 1$ is a nontrivial solution and the set S is linearly dependent. Moreover, from $-2\mathbf{v_1} - 3\mathbf{v_2} + \mathbf{v_3} = \theta$ we obtain $\mathbf{v_3} = 2\mathbf{v_1} + 3\mathbf{v_2}$. If $\mathbf{v}$ is in $\mathrm{Sp}(S)$ then $\mathbf{v} = a_1\mathbf{v_1} + a_2\mathbf{v_2} + a_3\mathbf{v_3} = (a_1 + 2a_3)\mathbf{v_1} + (a_2 + 3a_3)\mathbf{v_2}$, so $\mathbf{v}$ is in $\mathrm{Sp}\{\mathbf{v_1}, \mathbf{v_2}\}$. It follows that $\mathrm{Sp}\{\mathbf{v_1}, \mathbf{v_2}, \mathbf{v_3}\} = \mathrm{Sp}\{\mathbf{v_1}, \mathbf{v_2}\}$.

29. The subsets are $\{\mathbf{v_1}, \mathbf{v_2}, \mathbf{v_3}\}$, $\{\mathbf{v_1}, \mathbf{v_3}, \mathbf{v_4}\}$, and $\{\mathbf{v_1}, \mathbf{v_2}, \mathbf{v_4}\}$. Note that $\mathbf{v_4} = 3\mathbf{v_2} - \mathbf{v_3}$.

31. Set $V = [\mathbf{v_1}, \mathbf{v_2}, \mathbf{v_3}]$. By assumption the system $V\mathbf{x} = \mathbf{b}$ has a solution for every $\mathbf{b}$ in R^3. By Theorem 13 of Section 1.8, V is a nonsingular matrix. Therefore, by Theorem 12 of Section 1.8, the set $\{\mathbf{v_1}, \mathbf{v_2}, \mathbf{v_3}\}$ is linearly independent.

33. The set S is linearly dependent so S is not a basis for R^3.

35. If $\mathbf{u} = [u_1, u_2, u_3]^{\mathrm{T}}$ then $\mathbf{u}$ is in $\mathrm{Sp}(S)$ if and only if $4u_1 - 2u_2 + u_3 = 0$. In particular, $\mathrm{Sp}(S) \neq R^3$ and S is not a basis for R^3.

2.5 Dimension

1. S contains only one vector and $\dim(R^2) = 2$, so by property 2 of Theorem 9, S does not span R^2.

3. Since S contains three vectors and $\dim(R^2) = 2$, S is linearly dependent by property 1 of Theorem 9.

5. Since $\mathbf{u_4} = \theta$, S is a linearly dependent set; for example $0\mathbf{u_1} + a\mathbf{u_4} = \theta$ for any nonzero scalar a. Also S does not span R^2 since $\mathrm{Sp}\{\mathbf{u_1}, \mathbf{u_4}\} = \mathrm{Sp}\{\mathbf{u_1}\}$.

7. S contains two vectors and $\dim(R^3) = 3$ so by property 2 of Theorem 9, S does not span R^3.

9. Since S contains four vectors and $\dim(R^3) = 3$, S is linearly dependent by property 1 of Theorem 9.

11. It is easily checked that S is a linearly independent set. Since S contains two vectors and $\dim(R^2) = 2$ it follows from property 3 of Theorem 9 that S is a basis for R^2.

13. It is easily shown by direct calculation that S is a linearly dependent set. Therefore S is not a basis for R^3.

15. If we write $x_1 = 2x_2 - x_3 + x_4$ then the procedure described in Example 5 of Section 2.4 yields a basis $\{[2,1,0,0]^T, [-1,0,1,0]^T, [1,0,0,1]^T\}$ for W. It follows that $\dim(W) = 3$.

17. Following the procedure used in Example 5 of Section 2.4, we obtain a basis $\{[1,-1,0,0]^T, [2,0,-1,0]^T\}$ for W. In particular $\dim(W) = 2$.

19. The set $\{[-1,3,2,1]^T\}$ is a basis for W, so $\dim(W) = 1$.

21. The homogeneous system $A\mathbf{x} = \theta$ has solution $x_1 = -2x_2$. Therefore $\{[-2,1]^T\}$ is a basis for $\mathcal{N}(A)$ and $\text{nullity}(A) = 1$. Since $2 = \text{rank}(A) + \text{nullity}(A)$, it follows that $\text{rank}(A) = 1$.

23. The homogeneous system $A\mathbf{x} = \theta$ has solution $x_1 = -5x_3, x_2 = -2x_3$. Thus $\{[-5,-2,1]^T\}$ is a basis for $\mathcal{N}(A)$ and $\text{nullity}(A) = 1$. Since $3 = \text{rank}(A) + \text{nullity}(A)$, it follows that $\text{rank}(A) = 2$.

25. A^T reduces to $B^T = \begin{bmatrix} 1 & -1 & 1 \\ 0 & 2 & 3 \\ 0 & 0 & 0 \end{bmatrix}$. It follows that $\{[1,-1,1]^T, [0,2,3]^T\}$ is a basis for $\mathcal{R}(A)$. Consequently $\text{rank}(A) = 2$. Since $3 = \text{rank}(A) + \text{nullity}(A)$, it follows that $\text{nullity}(A) = 1$.

27. (a) Following the methods of Example 7 in Section 2.4, let $A = \begin{bmatrix} 1 & -1 & 1 & 2 \\ 1 & -2 & 0 & -1 \\ -2 & 3 & -1 & 0 \end{bmatrix}$.

Then A^T reduces to $B^T = \begin{bmatrix} 1 & 1 & -2 \\ 0 & -1 & 1 \\ 0 & 0 & 1 \\ 0 & 0 & 0 \end{bmatrix}$. It follows that $\{ [1,1,-2]^T, [0,-1,1]^T, [0,0,1]^T \}$ is a basis for W. In particular $\dim(W) = 3$.

(b) Following the procedure in (a), we obtain a basis $\{[1,2,-1,1]^T, [0,1,-1,1]^T, [0,0,-1,4]^T\}$ for W. In particular, $\dim(W) = 3$.

29. The constraints $a^Tx = 0, b^Tx = 0$ and $c^Tx = 0$ yield the homogeneous system of equations $x_1 - x_2 = 0$, $x_1 - x_3 = 0$, and $x_2 - x_3 = 0$. Solving we obtain $x_1 = x_3$ and $x_2 = x_3$ where x_3 and x_4 are arbitrary. Thus $\{[1,1,1,0]^T, [0,0,0,1]^T\}$ is a basis for W and $\dim(W) = 2$.

31. Suppose $\mathbf{x} = a_1\mathbf{u_1} + a_2\mathbf{u_2} + \cdots + a_p\mathbf{u_p}$ and $\mathbf{x} = b_1\mathbf{u_1} + b_2\mathbf{u_2} + \cdots + b_p\mathbf{u_p}$. Then $\theta = \mathbf{x} - \mathbf{x} = (a_1 - b_1)\mathbf{u_1} + (a_2 - b_2)\mathbf{u_2} + \cdots + (a_p - b_p)\mathbf{u_p}$. Since $\{\mathbf{u_1}, \mathbf{u_2}, \ldots, \mathbf{u_p}\}$ is linearly independent, $a_1 - b_1 = 0, a_2 - b_2 = 0, \ldots, a_p - b_p = 0$. Therefore $a_1 = b_1, a_2 = b_2, \ldots, a_p = b_p$.

33. (a) $\text{rank}(A) \le 3$ and $\text{nullity}(A) \ge 0$.

(b) $\text{rank}(A) \le 3$ and $\text{nullity}(A) \ge 1$.

(c) $\text{rank}(A) \le 4$ and $\text{nullity}(A) \ge 0$.

2.6 Orthogonal Bases for Subspaces

1. $\mathbf{u_1}^T\mathbf{u_2} = 1(-1) + 1(0) + 1(1) = 0; \mathbf{u_1}^T\mathbf{u_3} = 1(-1) + 1(2) + 1(-1) = 0; \mathbf{u_2}^T\mathbf{u_3} = -1(-1) + 0(2) + 1(-1) = 0$.

3. $\mathbf{u_1}^T\mathbf{u_2} = 1(2) + 1(0) + 2(-1) = 0; \mathbf{u_1}^T\mathbf{u_3} = 1(1) + 1(-5) + 2(2) = 0; \mathbf{u_2}^T\mathbf{u_3} = 2(1) + 0(-5) + (-1)2 = 0$.

5. $0 = \mathbf{u_1}^T\mathbf{u_3} = a + b + c$ and $0 = \mathbf{u_2}^T\mathbf{u_3} = 2a + 2b - 4c$. Solving yields $a = -b$, b arbitrary, and $c = 0$.

7. $0 = \mathbf{u_1}^T\mathbf{u_2} = -3 + a; 0 = \mathbf{u_1}^T\mathbf{u_3} = 4 + b + c; 0 = \mathbf{u_2}^T\mathbf{u_3} = -8 - b + ac$. Solving yields $a = 3$, $b = -5$, $c = 1$.

9. $\mathbf{v} = a_1\mathbf{u_1} + a_2\mathbf{u_2} + a_3\mathbf{u_3}$ where $a_1 = (\mathbf{u_1}^T\mathbf{v})/(\mathbf{u_1}^T\mathbf{u_1}) = 2/3$, $a_2 = (\mathbf{u_2}^T\mathbf{v})/(\mathbf{u_2}^T\mathbf{u_2}) = -1/2$, $a_3 = (\mathbf{u_3}^T\mathbf{v})/(\mathbf{u_3}^T\mathbf{u_3}) = 1/6$.

11. $\mathbf{v} = a_1\mathbf{u_1} + a_2\mathbf{u_2} + a_3\mathbf{u_3}$ where $a_1 = (\mathbf{u_1}^T\mathbf{v})/(\mathbf{u_1}^T\mathbf{u_1}) = 9/3 = 3$, $a_2 = (\mathbf{u_2}^T\mathbf{v})/(\mathbf{u_2}^T\mathbf{u_2}) = 0$, $a_3 = (\mathbf{u_3}^T\mathbf{v})/(\mathbf{u_3}^T\mathbf{u_3}) = 0$.

13. Denote the given vectors by, $\mathbf{w_1}$, $\mathbf{w_2}$, $\mathbf{w_3}$, respectively. Then $\mathbf{u_1} = \mathbf{w_1} = [0,0,1,0]^T$. $\mathbf{u_2} = \mathbf{w_2} - c_1\mathbf{u_1}$, where $c_1 =$

$(\mathbf{u_1}^T\mathbf{w_2})/(\mathbf{u_1}^T\mathbf{u_1}) = 2$. Then $\mathbf{u_2} = [1,1,0,1]^T$, $\mathbf{u_3} = \mathbf{w_3} - b_1\mathbf{u_1} - b_2\mathbf{u_2}$ where $b_1 = (\mathbf{u_1}^T\mathbf{w_3})/(\mathbf{u_1}^T\mathbf{u_1}) = 1$ and
$b_2 = (\mathbf{u_2}^T\mathbf{w_3})/(\mathbf{u_2}^T\mathbf{u_2}) = 2/3$. Therefore $\mathbf{u_3} = [1/3, -2/3, 0, 1/3]^T$

15. Denote the given vectors by $\mathbf{w_1}, \mathbf{w_2}, \mathbf{w_3}$, respectively. Then $\mathbf{u_1} = \mathbf{w_1} = [1,1,0]^T$, $\mathbf{u_2} = \mathbf{w_2} - c_1\mathbf{u_1}$, where $c_1 = (\mathbf{u_1}^T\mathbf{w_2})/(\mathbf{u_1}^T\mathbf{u_1}) = 2/2 = 1$. Thus $\mathbf{u_2} = [-1,1,1]^T$. $\mathbf{u_3} = \mathbf{w_3} - b_1\mathbf{u_1} - b_2\mathbf{u_2}$, where $b_1 = (\mathbf{u_1}^T\mathbf{w_3})/(\mathbf{u_1}^T\mathbf{u_1}) = 2/2 = 1$ and $b_2 = (\mathbf{u_2}^T\mathbf{w_3})/(\mathbf{u_2}^T\mathbf{u_2}) = 6/3 = 2$. Therefore $\mathbf{u_3} = [2, -2, 4]^T$.

17. Denote the given system by $\mathbf{w}_1$, $\mathbf{w}_2$, $\mathbf{w}_3$, respectively. Then $\mathbf{u}_1 = \mathbf{w}_1 = [0, 1, 0, 1]^T$, $\mathbf{u}_2 = \mathbf{w}_2 - c_1\mathbf{u}_1$, where $c_1 = (\mathbf{u}_1{}^T\mathbf{w}_2)/(\mathbf{u}_1{}^T\mathbf{u}_1) = 2/2 = 1$. Thus $\mathbf{u}_2 = [1, 1, 0, -1]^T$, $\mathbf{u}_3 = \mathbf{w}_3 - b_1\mathbf{u}_1 - b_2\mathbf{u}_2$, where $b_1 = (\mathbf{u}_1{}^T\mathbf{w}_3)/(\mathbf{u}_1{}^T\mathbf{u}_1) = 2/2 = 1$ and $b_2 = (\mathbf{u}_2{}^T\mathbf{w}_3)/(\mathbf{u}_2{}^T\mathbf{u}_2) = 2/3$. Therefore $\mathbf{u}_3 = [-2/3, 1/3, 1, -1/3]^T$.

19. If A denotes the given matrix then the homogeneous system $A\mathbf{x} = \theta$ has solution $x_1 = -3x_3 - x_4$, $x_2 = -x_3 - 3x_4$, where x_3 and x_4 are arbitrary. It follows that $\{[-3, -1, 1, 0]^T, [-1, -3, 0, 1]^T\}$ is a basis for $\mathcal{N}(A)$ and $\{\mathbf{A}_1, \mathbf{A}_2\}$ is a basis for $\mathcal{R}(A)$, where $\mathbf{A}_1 = [1, 2, 1]^T$ and $\mathbf{A}_2 = [-2, 1, -1]^T$. The Gram-Schmidt process yields orthogonal bases $\{[-3, -1, 1, 0]^T, [7/11, -27/11, -6/11, 1]^T\}$ and $\{[1, 2, 1]^T, [-11/6, 8/6, -5/6]^T\}$ for $\mathcal{N}(A)$ and $\mathcal{R}(A)$ respectively.

2.7 Linear Transformations from R^n to R^m

1. (a) $T\left(\begin{bmatrix} 0 \\ 0 \end{bmatrix}\right) = \begin{bmatrix} 0 \\ 0 \end{bmatrix}$.

   (b) $T\left(\begin{bmatrix} 1 \\ 1 \end{bmatrix}\right) = \begin{bmatrix} -1 \\ 0 \end{bmatrix}$.

   (c) $T\left(\begin{bmatrix} 2 \\ 1 \end{bmatrix}\right) = \begin{bmatrix} 1 \\ -1 \end{bmatrix}$.

(d) $T\left(\begin{bmatrix} -1 \\ 0 \end{bmatrix}\right) = \begin{bmatrix} -2 \\ 1 \end{bmatrix}.$

3. (a), (b), and (d) are in the null space of T.

5. If $\mathbf{b} = [b_1, b_2]^T$ then $T(\mathbf{x}) = \mathbf{b}$ requires that $2x_1 - 3x_2 = b_1$ and $-x_1 + x_2 = b_2$. Solving yields $x_1 = -b_1 - 3b_2$ and $x_2 = -b_1 - 2b_2$; that is $\mathbf{x} = [-b_1 - 3b_2, -b_1 - 2b_2]^T$.

7. The system of equations $A\mathbf{x} = \mathbf{b}$ is easily seen to be inconsistent.

9. F is a linear transformation.

11. F is not a linear transformation. For example $F\left(\begin{bmatrix} 1 \\ 2 \end{bmatrix} + \begin{bmatrix} 2 \\ 1 \end{bmatrix}\right) = F\left(\begin{bmatrix} 3 \\ 3 \end{bmatrix}\right) = \begin{bmatrix} 9 \\ 9 \end{bmatrix}$

whereas $F\left(\begin{bmatrix} 1 \\ 2 \end{bmatrix}\right) + F\left(\begin{bmatrix} 2 \\ 1 \end{bmatrix}\right) = \begin{bmatrix} 1 \\ 2 \end{bmatrix} + \begin{bmatrix} 4 \\ 2 \end{bmatrix} = \begin{bmatrix} 5 \\ 4 \end{bmatrix}.$

13. F is a linear transformation.

15. F is a linear transformation.

17. F is not a linear transformation. For example note that $F(-\mathbf{e}_1) = 1$ whereas $-F(\mathbf{e}_1) = -1$.

19. (a) $T\left(\begin{bmatrix} 1 \\ 1 \end{bmatrix}\right) = T(\mathbf{e}_1 + \mathbf{e}_2) = T(\mathbf{e}_1) + T(\mathbf{e}_2) = \mathbf{u}_1 + \mathbf{u}_2 = [3, 1, -1]^T.$

(b) $T\left(\begin{bmatrix} 2 \\ -1 \end{bmatrix}\right) = T(2\mathbf{e}_1 - \mathbf{e}_2) = 2T(\mathbf{e}_1) - T(\mathbf{e}_2) = 2\mathbf{u}_1 - \mathbf{u}_2 = [0, -1, -2]^T.$

(c) $T\left(\begin{bmatrix} 3 \\ 2 \end{bmatrix}\right) = T(3\mathbf{e}_1 + 2\mathbf{e}_2) = 3T(\mathbf{e}_1) + 2T(\mathbf{e}_2) = 3\mathbf{u}_1 + 2\mathbf{u}_2 = [7, 2, -3]^T.$

21. Let $\mathbf{u}_1 = [1, 1]^T$ and $\mathbf{u}_2 = [1, -1]^T$. If $\mathbf{x} = [x_1, x_2]^T$ then $\mathbf{x} = [(x_1 + x_2)/2]\mathbf{u}_1 + [(x_1 - x_2)/2]\mathbf{u}_2$. Thus $T(\mathbf{x}) =$

$[(x_1 + x_2)/2]\begin{bmatrix} 2 \\ -1 \end{bmatrix} + [(x_1 - x_2)/2]\begin{bmatrix} 0 \\ 3 \end{bmatrix} = \begin{bmatrix} x_1 + x_2 \\ x_1 - 2x_2 \end{bmatrix}.$

23. Let $\mathbf{u}_1 = [1, 0, 1]^T$, $\mathbf{u}_2 = [0, -1, 1]^T$, $\mathbf{u}_3 = [1, -1, 0]^T$. If $\mathbf{x} = [x_1, x_2, x_3]^T$ then $\mathbf{x} = c_1\mathbf{u}_1 + c_2\mathbf{u}_2 + c_3\mathbf{u}_3$, where $c_1 = (x_1 + x_2 + x_3)/2$, $c_2 = (-x_1 - x_2 + x_3)/2$ and $c_3 = (x_1 - x_2 - x_3)/2$.

Therefore $T(\mathbf{x}) = c_1[0, 1]^T + c_2[1, 0]^T + c_3[0, 0]^T$; that is,

$$T\left(\begin{bmatrix} x_1 \\ x_2 \\ x_3 \end{bmatrix}\right) = \begin{bmatrix} (-x_1 - x_2 + x_3)/2 \\ (x_1 + x_2 + x_3)/2 \end{bmatrix}.$$

25. $A = [T(\mathbf{e}_1), T(\mathbf{e}_2)] = \begin{bmatrix} 1 & 3 \\ 2 & 1 \end{bmatrix}$. The homogeneous system of equations $A\mathbf{x} = \theta$ has only the trivial solution so $\mathcal{N}(T) = \mathcal{N}(A) = \{\theta\}$ and nullity $(T) = 0$. Since rank $(T) = 2 -$ nullity $(T) = 2$, it follows that $\mathcal{R}(T) = R^2$.

27. $A = [T(\mathbf{e}_1), T(\mathbf{e}_2)] = [3, 2]; \mathcal{N}(T) = \{\mathbf{x} : 3x_1 + 2x_2 = 0\}; \mathcal{R}(T) = R^1$; rank $(T) = 1$; nullity $(T) = 1$.

rank $(T) = 3$; nullity $(T) = 0$.

29. $A = [T(\mathbf{e}_1), T(\mathbf{e}_2), T(\mathbf{e}_3)] = \begin{bmatrix} 1 & -1 & 0 \\ 0 & 1 & -1 \end{bmatrix}; \mathcal{N}(T) = \mathcal{N}(A) =$

$\{\mathbf{x} : x_1 = x_3, x_2 = x_3\}; \mathcal{R}(T) = \mathcal{R}(A) = R^2$; rank $(T) = 2$;

nullity $(T) = 1$.

31. For any x and y in R, $f(x + y) = a(x + y) = ax + ay = f(x) + f(y)$. If b is any real number then $f(bx) = a(bx) = b(ax) = bf(x)$. Therefore f is a linear transformation.

33. $T\left(\begin{bmatrix} x_1 \\ x_2 \end{bmatrix}\right) = \begin{bmatrix} x_1 \\ -x_2 \end{bmatrix}$

45. (a) $A = \begin{bmatrix} 0 & -1 \\ 1 & 0 \end{bmatrix}$.

(b) $A = \begin{bmatrix} 1/2 & -\sqrt{3}/2 \\ \sqrt{3}/2 & 1/2 \end{bmatrix}$.

(c) $A = \begin{bmatrix} -1/2 & -\sqrt{3}/2 \\ \sqrt{3}/2 & -1/2 \end{bmatrix}$.

2.8 Least-Squares Solutions to Inconsistent Systems

1. $A^TA = \begin{bmatrix} 3 & 4 \\ 4 & 14 \end{bmatrix}$ and $A^T\mathbf{b} = \begin{bmatrix} 1 \\ 6 \end{bmatrix}$. The system of equations $A^TA\mathbf{x} = A^T\mathbf{b}$ has unique solution $\mathbf{x}^* = \begin{bmatrix} -5/13 \\ 7/13 \end{bmatrix}$.

3. $A^TA\mathbf{x} = \begin{bmatrix} 11 & 16 & 17 \\ 16 & 30 & 18 \\ 17 & 18 & 33 \end{bmatrix}$ and $A^T\mathbf{b} = \begin{bmatrix} 10 \\ 17 \\ 13 \end{bmatrix}$. The system of equations $A^TA\mathbf{x} = A^T\mathbf{b}$ has solution $\mathbf{x}^* = \begin{bmatrix} 28/74 \\ 27/74 \\ 0 \end{bmatrix} + x_3 \begin{bmatrix} -3 \\ 1 \\ 1 \end{bmatrix}$ where x_3 is arbitrary.

5. $A^TA = \begin{bmatrix} 14 & 28 \\ 28 & 56 \end{bmatrix}$ and $A^T\mathbf{b} = \begin{bmatrix} 52 \\ 104 \end{bmatrix}$. The system of equations $A^TA\mathbf{x} = A^T\mathbf{b}$ has solution $\mathbf{x}^* = \begin{bmatrix} 26/7 \\ 0 \end{bmatrix} + x_2 \begin{bmatrix} -2 \\ 1 \end{bmatrix}$, where x_2 is arbitrary.

7. We must obtain the least-squares solution to $A\mathbf{x} = \mathbf{b}$ where $A = \begin{bmatrix} -1 & 1 \\ 0 & 1 \\ 1 & 1 \\ 2 & 1 \end{bmatrix}$, $\mathbf{x} = \begin{bmatrix} m \\ c \end{bmatrix}$, and $\mathbf{b} = \begin{bmatrix} 0 \\ 1 \\ 2 \\ 4 \end{bmatrix}$. $A^TA = \begin{bmatrix} 6 & 2 \\ 2 & 4 \end{bmatrix}$ and

$A^T\mathbf{b} = \begin{bmatrix} 10 \\ 7 \end{bmatrix}$. The system of equations $A^TA\mathbf{x} = A^Tb$ has

solution $\mathbf{x}^* = \begin{bmatrix} 1.3 \\ 1.1 \end{bmatrix}$. Therefore $y = (1.3)t + 1.1$ is the least-squares linear fit.

9. We must obtain the least-squares solution to $A\mathbf{x} = \mathbf{b}$ where $A = \begin{bmatrix} -1 & 1 \\ 0 & 1 \\ 1 & 1 \\ 2 & 1 \end{bmatrix}$, $\mathbf{x} = \begin{bmatrix} m \\ c \end{bmatrix}$, and $\mathbf{b} = \begin{bmatrix} -1 \\ 1 \\ 3 \\ 5 \end{bmatrix}$. In this case $A^TA = \begin{bmatrix} 6 & 2 \\ 2 & 4 \end{bmatrix}$ and $A^T\mathbf{b} = \begin{bmatrix} 14 \\ 8 \end{bmatrix}$. The system of

equations $A^{\mathrm{T}}A\mathbf{x} = A^{\mathrm{T}}\mathbf{b}$ has solution $\mathbf{x}^* = [2,1]^{\mathrm{T}}$ so $y = 2t + 1$ is the least-squares linear fit.

11. We must obtain the least-squares solution to $A\mathbf{x} = \mathbf{b}$ where $A = \begin{bmatrix} 1 & -2 & 4 \\ 1 & -1 & 1 \\ 1 & 1 & 1 \\ 1 & 2 & 4 \end{bmatrix}$, $\mathbf{x} = \begin{bmatrix} a_0 \\ a_1 \\ a_2 \end{bmatrix}$,

and $\mathbf{b} = \begin{bmatrix} 2 \\ 1 \\ 1 \\ 2 \end{bmatrix}$. In this case $A^{\mathrm{T}}A =$

$\begin{bmatrix} 4 & 0 & 10 \\ 0 & 10 & 0 \\ 10 & 0 & 34 \end{bmatrix}$ and $A^{\mathrm{T}}\mathbf{b} = \begin{bmatrix} 6 \\ 0 \\ 18 \end{bmatrix}$. The system of equations

$A^{\mathrm{T}}A\mathbf{x} = A^{\mathrm{T}}\mathbf{b}$ has solution $\mathbf{x}^* = [2/3, 0, 1/3]^{\mathrm{T}}$ so $y = a_0 + a_1 t + a_2 t^2 = 2/3 + (1/3)t^2$ is the least-squares quadratic fit.

13. We must obtain the least-squares solution to $A\mathbf{x} = \mathbf{b}$ where $A = \begin{bmatrix} 1 & -2 & 4 \\ 1 & -1 & 1 \\ 1 & 0 & 0 \\ 1 & 1 & 1 \end{bmatrix}$, $\mathbf{x} = \begin{bmatrix} a_0 \\ a_1 \\ a_2 \end{bmatrix}$,

and $\mathbf{b} = \begin{bmatrix} -3 \\ -1 \\ 1 \\ 3 \end{bmatrix}$. In this case $A^{\mathrm{T}}A =$

$\begin{bmatrix} 4 & -2 & 6 \\ -2 & 6 & -8 \\ 6 & -8 & 18 \end{bmatrix}$ and $A^{\mathrm{T}}\mathbf{b} = \begin{bmatrix} 0 \\ 10 \\ -10 \end{bmatrix}$. The system of equations

$A^{\mathrm{T}}A\mathbf{x} = A^{\mathrm{T}}\mathbf{b}$ has solution $\mathbf{x}^* = [1, 2, 0]^{\mathrm{T}}$ so $y = 1 + 2t$ is the least-squares quadratic fit.

2.9 Fitting Data and Least Squares Solutions

1. If $\mathbf{u_1} = [2,1,0]^{\mathrm{T}}$ and $\mathbf{u_2} = [-1,0,1]^{\mathrm{T}}$ then $\{\mathbf{u_1}, \mathbf{u_2}\}$ is a basis for W. For $A = [\mathbf{u_1}, \mathbf{u_2}]$ the system of equations $A^{\mathrm{T}}A\mathbf{x} = A^{\mathrm{T}}\mathbf{v}$ is given by $5x_1 - 2x_2 = 4$, $-2x_1 + 2x_2 = 5$. Solving we obtain $x_1 = 3$ and $x_2 = 11/2$. Thus $\mathbf{w}^* = 3\mathbf{u_1} + (11/2)\mathbf{u_2} = [1/2, 3, 11/2]^{\mathrm{T}}$.

3. The basis $\{\mathbf{u_1}, \mathbf{u_2}\}$ is given in Exercise 1. The system $A^{\mathrm{T}}A\mathbf{x} = A^{\mathrm{T}}\mathbf{v}$ is given by $5x_1 - 2x_2 = 3$, $-2x_1 + 2x_2 = 0$. The solution is $x_1 = x_2 = 1$, so $\mathbf{w}^* = \mathbf{u_1} + \mathbf{u_2} = [1,1,1]^{\mathrm{T}} = \mathbf{v}$

5. $\mathcal{R}(B)$ has basis $\{\mathbf{u_1}, \mathbf{u_2}\}$ where $\mathbf{u_1} = [1,1,0]^{\mathrm{T}}$ and $\mathbf{u_2} = [2,1,1]^{\mathrm{T}}$. If $A = [\mathbf{u_1}, \mathbf{u_2}]$, the system $A^{\mathrm{T}}A\mathbf{x} = A^{\mathrm{T}}\mathbf{v}$ is given by $2x_1 + 3x_2 = 6$, $3x_1 + 6x_2 = 12$. Solving yields $x_1 = 0$, $x_2 = 2$, so $\mathbf{w}^* = 2\mathbf{u_2} = [4,2,2]^{\mathrm{T}}$.

7. $\mathcal{R}(B)$ has basis $\{\mathbf{u_1}, \mathbf{u_2}\}$ where $\mathbf{u_1} = [1,-1,1]^{\mathrm{T}}$ and $\mathbf{u_2} = [2,0,1]^{\mathrm{T}}$. If $A = [\mathbf{u_1}, \mathbf{u_2}]$, the system $A^{\mathrm{T}}A\mathbf{x} = A^{\mathrm{T}}\mathbf{v}$ is given by $3x_1 + 3x_2 = 6$, $3x_1 + 5x_2 = 8$. Solving yields $x_1 = x_2 = 1$, so $\mathbf{w}^* = \mathbf{u_1} + \mathbf{u_2} = [3,-1,2]^{\mathrm{T}}$.

9. W has basis $\{\mathbf{u}\}$ where $\mathbf{u} = [0,-1,1]^{\mathrm{T}}$. If $A = [\mathbf{u}]$ then the system $A^{\mathrm{T}}A\mathbf{x} = A^{\mathrm{T}}\mathbf{v}$ is given by $2x = -2$. Thus $x = -1$ and $\mathbf{w}^* = -\mathbf{u} = [0,1,-1]^{\mathrm{T}}$.

11. An orthogonal basis for W is the set $\{\mathbf{u_1}, \mathbf{u_2}\}$ where $\mathbf{u_1} = [2,1,0]^{\mathrm{T}}$ and $\mathbf{u_2} = [-1/5, 2/5, 1]^{\mathrm{T}}$. The vector $\mathbf{w}^*$ is given by $\mathbf{w}^* = a_1\mathbf{u_1} + a_2\mathbf{u_2}$ where $a_1 = \mathbf{u_1}^{\mathrm{T}}\mathbf{v}/\mathbf{u_1}^{\mathrm{T}}\mathbf{u_1} = 4/5$ and $a_2 = \mathbf{u_2}^{\mathrm{T}}\mathbf{v}/\mathbf{u_2}^{\mathrm{T}}\mathbf{u_2} = 11/2$. Thus $\mathbf{w}^* = [1/2, 3, 11/2]^{\mathrm{T}}$.

13. If $\mathbf{u_1} = [1,1,0]^{\mathrm{T}}$ and $\mathbf{u_2} = [1/2,-1/2,1]^{\mathrm{T}}$ then $\{\mathbf{u_1}, \mathbf{u_2}\}$ is an orthogonal basis for W. The vector $\mathbf{w}^*$ is given by $\mathbf{w}^* = a_1\mathbf{u_1} + a_2\mathbf{u_2}$ where $a_1 = \mathbf{u_1}^{\mathrm{T}}\mathbf{v}/\mathbf{u_1}^{\mathrm{T}}\mathbf{u_1} = 1$ and $a_2 = \mathbf{u_2}^{\mathrm{T}}\mathbf{v}/\mathbf{u_2}^{\mathrm{T}}\mathbf{u_2} = 4$. Thus $\mathbf{w}^* = [3,-1,4]^{\mathrm{T}}$.

15. If $\mathbf{u_1} = [1,-1,1]^{\mathrm{T}}$ and $\mathbf{u_2} = [1,1,0]^{\mathrm{T}}$ then $\{\mathbf{u_1}, \mathbf{u_2}\}$ is an orthogonal basis for W. The vector $\mathbf{w}^*$ is given by $\mathbf{w}^* = a_1\mathbf{u_1} + a_2\mathbf{u_2}$ where $a_1 = \mathbf{u_1}^{\mathrm{T}}\mathbf{v}/\mathbf{u_1}^{\mathrm{T}}\mathbf{u_1} = 2$ and $a_2 = \mathbf{u_2}^{\mathrm{T}}\mathbf{v}/\mathbf{u_2}^{\mathrm{T}}\mathbf{u_2} = 1$. Therefore $\mathbf{w}^* = [3,-1,2]^{\mathrm{T}}$.

$\mathcal{R}(A)$ has basis $\{u_1, u_2\}$ where $u_1 = [1,1,0]^T$ and $u_2 = [2,0,1]^T$. If $A = [u_1, u_2]$, the system $A^TAx = A^Tv$ is given by $2x_1 + [illegible]x_2 = 6$, $2x_1 + 5x_2 = 12$. Solving yields $x_1 = [illegible]$, $x_2 = 2$, so $w^* = [illegible]2u_2 = [4, [illegible], 2]^T$.

$\mathcal{R}(B)$ has basis $\{u_1, u_2\}$ where $u_1 = [1, [illegible], 1]^T$ and $u_2 = [2, 0, [illegible]]^T$. If $A = [u_1, u_2]$, the system $A^TAx = A^Tv$ is given by $3x_1 + [illegible]x_2 = 6$, $[illegible]x_1 + [illegible]x_2 = [illegible]$. Solving yields $x_1 = [illegible]$, $[illegible]$, so $w^* = u_1 + u_2 = [3, -1, 2]^T$.

W has basis $\{u\}$ where $u = [0, [illegible], 1]^T$. If $A = [u]$, then the system $A^TAx = A^Tv$ is $[illegible]x = -2$. Thus $x = -1$ and $w^* = -u = [0, [illegible], -1]^T$.

An orthogonal basis for W is the set $\{u_1, u_2\}$, where $u_1 = [2, 1, 0]^T$ and $u_2 = [-1/5, 2/5, 1]^T$. The vector w^* is given by $w^* = a_1u_1 + a_2u_2$, where $a_1 = u_1^Tv / u_1^Tu_1 = 4/5$ and $a_2 = u_2^Tv / u_2^Tu_2 = 11/2$. Thus $w^* = [1/2, 3, 11/2]^T$.

If $u_1 = [1, 1, 0]^T$ and $u_2 = [1/2, -1/2, 1]^T$, then $\{u_1, u_2\}$ is an orthogonal basis for W. The vector w^* is given by $w^* = a_1u_1 + a_2u_2$, where $a_1 = u_1^Tv / u_1^Tu_1 = 1$ and $a_2 = u_2^Tv / u_2^Tu_2 = [illegible]$. Thus $w^* = [illegible]$.

If $u_1 = [1, -1, 1]^T$ and $u_2 = [1, 1, 0]^T$, then $\{u_1, u_2\}$ is an orthogonal basis for W. The vector w^* is given by $w^* = a_1u_1 + a_2u_2$, where $a_1 = u_1^Tv / u_1^Tu_1 = 2$ and $a_2 = u_2^Tv / u_2^Tu_2 = 1$. Therefore $w^* = [3, -1, 2]^T$.

Chapter 3

The Eigenvalue Problems

3.1 Introduction

1. The matrix $A - \lambda I = \begin{bmatrix} 1-\lambda & 0 \\ 2 & 3-\lambda \end{bmatrix}$ is singular if and only if $0 = (1-\lambda)(3-\lambda)$. Thus $\lambda = 1$ and $\lambda = 3$ are eigenvalues for A. The eigenvectors corresponding to $\lambda = 1$ $(A - I)\mathbf{x} = \theta$. Solving yields $x_1 = -x_2$, x_2 arbitrary. Therefore any vector of the form $\mathbf{x} = a\begin{bmatrix} -1 \\ 1 \end{bmatrix}, a \neq 0,$ is an eigenvector for $\lambda = 1$. Similarly the eigenvectors corresponding to $\lambda = 3$ are the nontrivial solutions to $(A - 3I)\mathbf{x} = \theta$. Solving yields $x_1 = 0, x_2$ arbitrary, so any vector of the form $\mathbf{x} = a\begin{bmatrix} 0 \\ 1 \end{bmatrix}, a \neq 0,$ is an eigenvector for $\lambda = 3$.

3. The matrix $A - \lambda I = \begin{bmatrix} 2-\lambda & -1 \\ -1 & 2-\lambda \end{bmatrix}$ is singular if and only if $0 = (2-\lambda)(2-\lambda) - 1 = \lambda^2 - 4\lambda + 3 = (\lambda - 1)(\lambda - 3)$. Therefore A has eigenvalues $\lambda = 1$ and $\lambda = 3$. Solving $(A - I)\mathbf{x} = \theta$ yields $x_1 = x_2, x_2$ arbitrary, so any vector of the form $\mathbf{x} = a\begin{bmatrix} 1 \\ 1 \end{bmatrix}, a \neq 0,$ is an eigenvector corresponding to $\lambda = 1$. Solving $(A - 3I)\mathbf{x} = \theta$ yields $x_1 = -x_2, x_2$ arbitrary, so any vector of the form $\mathbf{x} = a\begin{bmatrix} -1 \\ 1 \end{bmatrix}, a \neq 0,$ is an eigenvector for $\lambda = 3$.

5. The matrix $A - \lambda I = \begin{bmatrix} 2-\lambda & 1 \\ 1 & 2-\lambda \end{bmatrix}$ is singular if and only if $0 = (2-\lambda)(2-\lambda) - 1 = \lambda^2 - 4\lambda + 3 = (\lambda - 1)(\lambda - 3)$. Therefore A has eigenvalues $\lambda = 1$ and $\lambda = 3$. Solving $(A - I)\mathbf{x} = \theta$ yields $x_1 = -x_2, x_2$ arbitrary, so any vector of the form $\mathbf{x} = a\begin{bmatrix} -1 \\ 1 \end{bmatrix}, a \neq 0,$ is an eigenvector for $\lambda = 1$. Solving $(A - 3I)\mathbf{x} = \theta$ yields $x_1 = x_2, x_2$ arbitrary, so any

vector $\mathbf{x} = a \begin{bmatrix} 1 \\ 1 \end{bmatrix}$, $a \neq 0$, is an eigenvector for $\lambda = 3$.

7. The matrix $A - \lambda I = \begin{bmatrix} 1-\lambda & 0 \\ 2 & 1-\lambda \end{bmatrix}$ is singular if and only if $0 = (1-\lambda)^2$, so $\lambda = 1$ is the only eigenvalue for A. Solving $(A - I)\mathbf{x} = \theta$ yields $x_1 = 0, x_2$ arbitrary, so any vector $\mathbf{x} = a \begin{bmatrix} 0 \\ 1 \end{bmatrix}$, $a \neq 0$, is an eigenvector for $\lambda = 1$.

9. The matrix $A - \lambda I = \begin{bmatrix} 2-\lambda & 2 \\ 3 & 3-\lambda \end{bmatrix}$ is singular if and only if $0 = (2-\lambda)(3-\lambda) - 6 = \lambda^2 - 5\lambda = \lambda(\lambda - 5)$. Therefore A has eigenvalues $\lambda = 0$ and $\lambda = 5$. Solving $A\mathbf{x} = \theta$ yields $x_1 = -x_2, x_2$ arbitrary, so $\mathbf{x} = a \begin{bmatrix} -1 \\ 1 \end{bmatrix}$, $a \neq 0$, is an eigenvector for $\lambda = 0$. Solving $(A - 5I)\mathbf{x} = \theta$ yields $x_1 = (2/3)x_2, x_2$ arbitrary, so $\mathbf{x} = a \begin{bmatrix} 2 \\ 3 \end{bmatrix}$, $a \neq 0$, is an eigenvector for $\lambda = 5$.

11. The matrix $A - \lambda I = \begin{bmatrix} 1-\lambda & -1 \\ 1 & 3-\lambda \end{bmatrix}$ is singular if and only if $0 = (1-\lambda)(3-\lambda) + 1 = \lambda^2 - 4\lambda + 4 = (\lambda - 2)^2$. Therefore $\lambda = 2$ is the only eigenvalue for A. Solving $(A - 2I)\mathbf{x} = \theta$ yields $x_1 = -x_2$, so $\mathbf{x} = a \begin{bmatrix} -1 \\ 1 \end{bmatrix}$, $a \neq 0$ is an eigenvector for $\lambda = 2$.

13. The matrix $A - \lambda I = \begin{bmatrix} -2-\lambda & -1 \\ 5 & 2-\lambda \end{bmatrix}$ is singular if and only if $0 = (-2-\lambda)(2-\lambda) + 5 = \lambda^2 + 1$. Solving yields $\lambda = \pm i$.

15. The matrix $A - \lambda I = \begin{bmatrix} 2-\lambda & -1 \\ 1 & 2-\lambda \end{bmatrix}$ is singular if and only if $0 = (2-\lambda)(2-\lambda) + 1 = \lambda^2 - 4\lambda + 5$. Solving we obtain $\lambda = 2 \pm i$.

3.2 Determinants and the Eigenvalue Problem

1. $M_{11} = \begin{bmatrix} 1 & 3 & -1 \\ 2 & 4 & 1 \\ 2 & 0 & -2 \end{bmatrix}$. $A_{11} = \det(M_{11}) =$

$$\begin{vmatrix} 4 & 1 \\ 0 & -2 \end{vmatrix} - 3 \begin{vmatrix} 2 & 1 \\ 2 & -2 \end{vmatrix} - \begin{vmatrix} 2 & 4 \\ 2 & 0 \end{vmatrix} = 18.$$

3. $M_{31} = \begin{bmatrix} -1 & 3 & 1 \\ 1 & 3 & -1 \\ 2 & 0 & -2 \end{bmatrix}$. $A_{31} = \det(M_{31}) =$

$$-\begin{vmatrix} 3 & -1 \\ 0 & -2 \end{vmatrix} - 3\begin{vmatrix} 1 & -1 \\ 2 & -2 \end{vmatrix} + \begin{vmatrix} 1 & 3 \\ 2 & 0 \end{vmatrix} = 0.$$

5. $M_{34} = \begin{bmatrix} 2 & -1 & 3 \\ 4 & 1 & 3 \\ 2 & 2 & 0 \end{bmatrix}$. $A_{34} = -\det(M_{34}) =$

$$-2\begin{vmatrix} 1 & 3 \\ 2 & 0 \end{vmatrix} - \begin{vmatrix} 4 & 3 \\ 2 & 0 \end{vmatrix} - 3\begin{vmatrix} 4 & 1 \\ 2 & 2 \end{vmatrix} = 0.$$

7. $\det(A) = 2A_{11} + 4A_{21} + 6A_{31} + 2A_{41} =$
$2(18) + 4(-18) + 6(0) + 2(18) = 0.$

9. $\det(A) = 0; A$ is singular

11. $\det(A) = -1; A$ is nonsingular.

13. $\det(A) = 2\begin{vmatrix} -2 & 1 \\ 1 & -1 \end{vmatrix} + 3\begin{vmatrix} -1 & 1 \\ 3 & -1 \end{vmatrix} + 2\begin{vmatrix} -1 & -2 \\ 3 & 1 \end{vmatrix} = 6;$ A is nonsingular.

15. $\det(A) = 2\begin{vmatrix} 3 & 2 \\ 1 & 4 \end{vmatrix} = 20;$ A is nonsingular.

17. Expansion along the first column of A yields $\det(A) = \begin{vmatrix} 3 & 0 & 0 \\ 4 & 1 & 2 \\ 3 & 1 & 4 \end{vmatrix}$. Now expansion along

the first row gives $\det(A) = 3\begin{vmatrix} 1 & 2 \\ 1 & 4 \end{vmatrix} = 6.$

A is nonsingular.

19. Expansion along the first column in successive steps yields

$$det(A) = -3\begin{vmatrix} 0 & 0 & 2 \\ 0 & 3 & 1 \\ 2 & 1 & 2 \end{vmatrix} = (-3)(2)\begin{vmatrix} 0 & 2 \\ 3 & 1 \end{vmatrix} = (-3)(2)(-6) = 36.$$ A is nonsingular.

21. Det $(A) = 4x - 2y - 2$, so A is singular when $4x - 2y - 2 = 0$, that is when $y = 2x - 1$.

23. For $n = 2$, $\begin{vmatrix} d & 1 \\ 1 & d \end{vmatrix} = d^2 - 1 = (d-1)(d+1)$. For $n = 3$, $\begin{vmatrix} d & 1 & 1 \\ 1 & d & 1 \\ 1 & 1 & d \end{vmatrix} = d\begin{vmatrix} d & 1 \\ 1 & d \end{vmatrix} - \begin{vmatrix} 1 & 1 \\ 1 & d \end{vmatrix} +$

$$\begin{vmatrix} 1 & d \\ 1 & 1 \end{vmatrix} = d(d-1)(d+1) - (d-1) + (1-d)$$

$= (d-1)^2(d+2)$. For $n = 4$, $\begin{vmatrix} d & 1 & 1 & 1 \\ 1 & d & 1 & 1 \\ 1 & 1 & d & 1 \\ 1 & 1 & 1 & d \end{vmatrix} = d \begin{vmatrix} d & 1 & 1 \\ 1 & d & 1 \\ 1 & 1 & d \end{vmatrix} - \begin{vmatrix} 1 & 1 & 1 \\ 1 & d & 1 \\ 1 & 1 & d \end{vmatrix} + \begin{vmatrix} 1 & d & 1 \\ 1 & 1 & 1 \\ 1 & 1 & d \end{vmatrix} -$

$\begin{vmatrix} 1 & d & 1 \\ 1 & 1 & d \\ 1 & 1 & 1 \end{vmatrix} = d(d-1)^2(d+2) - 3(d-1)^2 = (d-1)^3(d+3).$

27. Det $(ABA^{-1}) = \det(A)\det(B)/\det(A) = \det(B) = 5$.

29. Det $(A^{-1}B^{-1}A^2) = [\det(A)]^2/[\det(A)\det(B)] = \det(A)/\det(B) = 3/5$.

31. (a) $H(n) = n!/2$.

(b) $n = 2$, 3 secs; $n = 5$, 3 min; $n = 10$, 63 days.

3.3 Elementary Operations and Determinants

1. $\det(A) = \begin{vmatrix} 1 & 2 & 1 \\ 3 & 0 & 2 \\ -1 & 1 & 3 \end{vmatrix} \begin{Bmatrix} R_2 - 3R_1 \\ R_3 + R_1 \end{Bmatrix} = \begin{vmatrix} 1 & 2 & 1 \\ 0 & -6 & -1 \\ 0 & 3 & 4 \end{vmatrix} =$

$\begin{vmatrix} -6 & -1 \\ 3 & 4 \end{vmatrix} = -21.$

3. $\det(A) = \begin{vmatrix} 3 & 6 & 9 \\ 2 & 0 & 2 \\ 1 & 2 & 0 \end{vmatrix} = (3)(2) \begin{vmatrix} 1 & 2 & 3 \\ 1 & 0 & 1 \\ 1 & 2 & 0 \end{vmatrix} \begin{Bmatrix} R_2 - R_1 \\ R_3 - R_1 \end{Bmatrix} =$

$(6) \begin{vmatrix} 1 & 2 & 3 \\ 0 & -2 & -2 \\ 0 & 0 & -3 \end{vmatrix} = 36.$

5. $\det(A) = \begin{vmatrix} 2 & 4 & -3 \\ 3 & 2 & 5 \\ 2 & 3 & 4 \end{vmatrix} = (1/2) \begin{vmatrix} 2 & 4 & -3 \\ 6 & 4 & 10 \\ 2 & 3 & 4 \end{vmatrix} \begin{Bmatrix} R_2 - 3R_1 \\ R_3 - R_1 \end{Bmatrix} =$

$(1/2) \begin{vmatrix} 2 & 4 & -3 \\ 0 & -8 & 19 \\ 0 & -1 & 7 \end{vmatrix} = (2)(1/2) \begin{vmatrix} -8 & 19 \\ -1 & 7 \end{vmatrix} = -37.$

7. $\begin{vmatrix} 1 & 0 & 0 & 0 \\ 2 & 0 & 0 & 3 \\ 1 & 1 & 0 & 1 \\ 1 & 4 & 2 & 2 \end{vmatrix} \{C_3 \leftrightarrow C_4\} = (-1) \begin{vmatrix} 1 & 0 & 0 & 0 \\ 2 & 0 & 3 & 0 \\ 1 & 1 & 1 & 0 \\ 1 & 4 & 2 & 2 \end{vmatrix} \{C_2 \leftrightarrow C_3\} =$

$$\begin{vmatrix} 1 & 0 & 0 & 0 \\ 2 & 3 & 0 & 0 \\ 1 & 1 & 1 & 0 \\ 1 & 2 & 4 & 2 \end{vmatrix} = (1)(3)(1)(2) = 6.$$

9. $$\begin{vmatrix} 0 & 0 & 2 & 0 \\ 0 & 0 & 1 & 3 \\ 0 & 4 & 1 & 3 \\ 2 & 1 & 5 & 6 \end{vmatrix} \underset{=}{\left\{ \begin{matrix} C_1 \leftrightarrow C_4 \\ C_2 \leftrightarrow C_3 \end{matrix} \right\}} \begin{vmatrix} 0 & 2 & 0 & 0 \\ 3 & 1 & 0 & 0 \\ 3 & 1 & 4 & 0 \\ 6 & 5 & 1 & 2 \end{vmatrix} \underset{=}{\{C_1 \leftrightarrow C_2\}} (-1)\begin{vmatrix} 2 & 0 & 0 & 0 \\ 1 & 3 & 0 & 0 \\ 1 & 3 & 4 & 0 \\ 5 & 6 & 1 & 2 \end{vmatrix} = (-1)(2)(3)(4)(2) = -48.$$

11. $$\begin{vmatrix} 0 & 0 & 1 & 0 \\ 0 & 2 & 6 & 3 \\ 2 & 4 & 1 & 5 \\ 0 & 0 & 0 & 4 \end{vmatrix} \underset{=}{\{R_1 \leftrightarrow R_3\}} (-1)\begin{vmatrix} 2 & 4 & 1 & 5 \\ 0 & 2 & 6 & 3 \\ 0 & 0 & 1 & 0 \\ 0 & 0 & 0 & 4 \end{vmatrix} = (-1)(2)(2)(1)(4) = -16.$$

13. $\det(B) = 3\det(A) = 6.$

15. $\det(B) = -\det(A) = -2.$

17. $\det(B) = -2\det(A) = -4.$

19. $$\begin{vmatrix} 2 & 4 & 2 & 6 \\ 1 & 3 & 2 & 1 \\ 2 & 1 & 2 & 3 \\ 1 & 2 & 1 & 1 \end{vmatrix} \underset{=}{\{R_1 - 2R_4\}} \begin{vmatrix} 0 & 0 & 0 & 4 \\ 1 & 3 & 2 & 1 \\ 2 & 1 & 2 & 3 \\ 1 & 2 & 1 & 1 \end{vmatrix} = (-4)\begin{vmatrix} 1 & 3 & 2 \\ 2 & 1 & 2 \\ 1 & 2 & 1 \end{vmatrix} \underset{=}{\left\{ \begin{matrix} R_2 - 2R_1 \\ R_3 - R_1 \end{matrix} \right\}} (-4)\begin{vmatrix} 1 & 3 & 2 \\ 0 & -5 & -2 \\ 0 & -1 & -1 \end{vmatrix} = (-4)\begin{vmatrix} -5 & -2 \\ -1 & -1 \end{vmatrix} = -12.$$

21. $$\begin{vmatrix} 0 & 4 & 1 & 3 \\ 0 & 2 & 2 & 1 \\ 1 & 3 & 1 & 2 \\ 2 & 2 & 1 & 4 \end{vmatrix} \underset{=}{\{R_4 - 2R_3\}} \begin{vmatrix} 0 & 4 & 1 & 3 \\ 0 & 2 & 2 & 1 \\ 1 & 3 & 1 & 2 \\ 0 & -4 & -1 & 0 \end{vmatrix} = \begin{vmatrix} 4 & 1 & 3 \\ 2 & 2 & 1 \\ -4 & -1 & 0 \end{vmatrix} \underset{=}{\{R_1 + R_3\}} \begin{vmatrix} 0 & 0 & 3 \\ 2 & 2 & 1 \\ -4 & -1 & 0 \end{vmatrix} = (3)\begin{vmatrix} 2 & 2 \\ -4 & -1 \end{vmatrix} = 18.$$

3.4 Eigenvalues and the Characteristic Polynomial

1. $p(t) = (1-t)(3-t)$. The eigenvalues are $\lambda = 1$ and $\lambda = 3$, each with algebraic multiplicity 1.

3. $p(t) = \begin{vmatrix} 2-t & -1 \\ -1 & 2-t \end{vmatrix} = (2-t)(2-t) - 1 = t^2 - 4t + 3 = (t-1)(t-3)$. The eigenvalues are $\lambda = 1$ånd $\lambda = 3$, each with algebraic multiplicity 1.

5. $p(t) = \begin{vmatrix} 1-t & -1 \\ 1 & 3-t \end{vmatrix} = (1-t)(3-t) + 1 = t^2 - 4t + 4 = (t-2)^2$. The only eigenvalue is $\lambda = 2$ and it has algebraic multiplicity 2.

7. $p(t) = \begin{vmatrix} -6-t & -1 & 2 \\ 3 & 2-t & 0 \\ -14 & -2 & 5-t \end{vmatrix} = -t^3 + t^2 + t - 1 = -(t-1)^2(t+1)$. The eigenvalues are $\lambda = 1$ with algebraic multiplicity 2 and $\lambda = -1$ with algebraic multiplicity 1.

9. $p(t) = \begin{vmatrix} 3-t & -1 & -1 \\ -12 & -t & 5 \\ 4 & -2 & -1-t \end{vmatrix} = -t^3 + 2t^2 + t - 2 = -(t-2)(t-1)(t+1)$. The eigenvalues are $\lambda = 2, \lambda = 1$,ånd $\lambda = -1$ each with algebraic multiplicity 1.

11. $p(t) = \begin{vmatrix} 2-t & 4 & 4 \\ 0 & 1-t & -1 \\ 0 & 1 & 3-t \end{vmatrix} = (2-t)\begin{vmatrix} 1-t & -1 \\ 1 & 3-t \end{vmatrix} = (2-t)(t^2 - 4t + 4) = -(t-2)^3.$
The only eigenvalue is $\lambda = 2$ and it has algebraic multiplicity 3.

13. $p(t) = \begin{vmatrix} 5-t & 4 & 1 & 1 \\ 4 & 5-t & 1 & 1 \\ 1 & 1 & 4-t & 2 \\ 1 & 1 & 2 & 4-t \end{vmatrix} = t^4 - 18t^3 + 97t^2 - 180t + 100 = (t-1)(t-2)(t-5)(t-10).$
The eigenvalues are $\lambda = 1, \lambda = 2, \lambda = 5, \lambda = 10$,èach with algebraic multiplicity 1.

19. $q(C) = C^3 - 2C^2 - C + 2I = \begin{bmatrix} 35 & -3 & -15 \\ -44 & 2 & 19 \\ 68 & -6 & -29 \end{bmatrix} -$
$2\begin{bmatrix} 17 & -1 & -7 \\ -16 & 2 & 7 \\ 32 & -2 & -13 \end{bmatrix} - \begin{bmatrix} 3 & -1 & -1 \\ -12 & 0 & 5 \\ 4 & -2 & -1 \end{bmatrix} + \begin{bmatrix} 2 & 0 & 0 \\ 0 & 2 & 0 \\ 0 & 0 & 2 \end{bmatrix} = \mathcal{O}.$

21. $p(t) = t^2 - 2t + 1$. $p(A) = A^2 - 2A + I = \mathcal{O}$.

23. $p(t) = t^4 - 18t^3 + 97t^2 - 180t + 100$. $p(A) = A^4 - 18A^3 + 97A^2 - 180A + 100I = \mathcal{O}$.

3.5 Eigenvalues and Eigenvectors

1. $(A - 3I)\mathbf{x} = \theta$ is the system

$$\begin{aligned} -x_1 - x_2 &= 0 \\ -x_1 - x_2 &= 0 \end{aligned}.$$

The solution is $x_1 = -x_2, x_2$ arbitrary, so E_λ consists of the vectors of the form $x_2 \begin{bmatrix} -1 \\ 1 \end{bmatrix}$. Thus $\{[-1, 1]^T\}$ is a basis for E_λ. The eigenvalue $\lambda = 3$ has algebraic and geometric multiplicity 1.

3. $(B - 2I)\mathbf{x} = \theta$ is the system

$$\begin{aligned} -x_1 - x_2 &= 0 \\ x_1 + x_2 &= 0 \end{aligned}.$$

The solution is $x_1 = -x_2, x_2$ arbitrary, so E_λ consists of the vectors of the form $x_2 \begin{bmatrix} -1 \\ 1 \end{bmatrix}$. Thus $\{[-1, 1]^T\}$ is a basis for E_λ. The eigenvalue $\lambda = 2$ has algebraic multiplicity 2 and geometric multiplicity 1.

5. $(C + I)\mathbf{x} = \theta$ is the system

$$\begin{array}{rcrcrcl} -5x_1 & - & x_2 & + & 2x_3 & = & 0 \\ 3x_1 & + & 3x_2 & & & = & 0 \\ -14x_1 & - & 2x_2 & + & 6x_3 & = & 0 \end{array}.$$

The solution is $x_1 = (1/2)x_3, x_2 = (-1/2)x_3, x_3$ arbitrary, so E_λ consists of vectors of the form $a \begin{bmatrix} 1 \\ -1 \\ 2 \end{bmatrix}$ where a is an arbitrary scalar. Thus $\{[1, -1, 2]^T\}$ is a basis for E_λ. The eigenvalue $\lambda = -1$ has algebraic and geometric multiplicity 1.

7. $(E + I)\mathbf{x} = \theta$ is the system

$$\begin{array}{rcrcrcrcl} 7x_1 & + & 4x_2 & + & 4x_3 & + & x_4 & = & 0 \\ 4x_1 & + & 7x_2 & + & x_3 & + & 4x_4 & = & 0 \\ 4x_1 & + & x_2 & + & 7x_3 & + & 4x_4 & = & 0 \\ x_1 & + & 4x_2 & + & 4x_3 & + & 7x_4 & = & 0 \end{array}.$$

The solution is $x_1 = x_4, x_2 = -x_4, x_3 = -x_4, x_4$ arbitary, so E_λ consists of vectors of the form $x_4 \begin{bmatrix} 1 \\ -1 \\ -1 \\ 1 \end{bmatrix}$. Thus $\{[1, -1, -1, 1]^T\}$

is a basis for E_λ.The eigenvalue $\lambda = -1$ has geometric and algebraic multiplicity 1.

9. $(E - 15I)\mathbf{x} = \theta$ is the system

$$\begin{array}{rcrcrcrcl} -9x_1 & + & 4x_2 & + & 4x_3 & + & x_4 & = & 0 \\ 4x_1 & - & 9x_2 & + & x_3 & + & 4x_4 & = & 0 \\ 4x_1 & + & x_2 & - & 9x_3 & + & 4x_4 & = & 0 \\ x_1 & + & 4x_2 & + & 4x_3 & - & 9x_4 & = & 0 \end{array}.$$

The solution is $x_1 = x_2 = x_3 = x_4, x_4$ arbitrary so E_λ consists of vectors of the form $x_4 \begin{bmatrix} 1 \\ 1 \\ 1 \\ 1 \end{bmatrix}$. Thus $\{[1,1,1,1]^{\mathrm{T}}\}$ is a basis for E_λ. The eigenvalue $\lambda = 15$ has algebraic and geometric multiplicity 1.

11. $(F - 2I)\mathbf{x} = \theta$ is the system

$$\begin{array}{rcrcrcrcl} -x_1 & - & x_2 & - & x_3 & - & x_4 & = & 0 \\ -x_1 & - & x_2 & - & x_3 & - & x_4 & = & 0 \\ -x_1 & - & x_2 & - & x_3 & - & x_4 & = & 0 \\ -x_1 & - & x_2 & - & x_3 & - & x_4 & = & 0 \end{array}.$$

The solution is $x_1 = -x_2 - x_3 - x_4, x_2, x_3, x_4$ arbitrary so E_λ consists of vectors of the form $x_2 \begin{bmatrix} -1 \\ 1 \\ 0 \\ 0 \end{bmatrix} + x_3 \begin{bmatrix} -1 \\ 0 \\ 1 \\ 0 \end{bmatrix} + x_4 \begin{bmatrix} -1 \\ 0 \\ 0 \\ 1 \end{bmatrix}$.Thus $\{[-1,1,0,0]^{\mathrm{T}}, [-1,0,1,0]^{\mathrm{T}}, [-1,0,0,1]^{\mathrm{T}}\}$ is a basis for E_λ.The eigenvalue $\lambda = 2$ has algebraic and geometric multiplicity 3.

13. The characteristic equation for the given matrix A is $0 = \det(A - tI) = \begin{vmatrix} 2-t & 1 & 2 \\ 0 & 3-t & 2 \\ 0 & 0 & 2-t \end{vmatrix} = -(t-2)^2(t-3)$.

The eigenvalues are $\lambda = 2$ and $\lambda = 3$.The system $(A - 2I)\mathbf{x} = \theta$ is given by

$$\begin{array}{l} x_2 + 2x_3 = 0 \\ x_2 + 2x_3 = 0 \end{array}.$$

In the solution x_1 is arbitrary, $x_2 = -2x_3$, and x_3 is arbitrary. The eigenvectors for $\lambda = 2$ are the nonzero vectors of the form $\begin{bmatrix} x_1 \\ -2x_3 \\ x_3 \end{bmatrix}$

$= x_1 \begin{bmatrix} 1 \\ 0 \\ 0 \end{bmatrix} + x_3 \begin{bmatrix} 0 \\ -2 \\ 1 \end{bmatrix}$. Therefore $\lambda = 2$ has algebraic and geometric multiplicity 2. The system $(A - 3I)\mathbf{x} = \theta$ is given by

$$\begin{aligned} -x_1 + x_2 + 2x_3 &= 0 \\ 2x_3 &= 0\,. \\ -x_3 &= 0 \end{aligned}$$

The solution is $x_1 = x_2, x_2$ arbitrary, $x_3 = 0$.The eigenvectors for $\lambda = 3$ are the nonzero vectors of the form $\begin{bmatrix} x_2 \\ x_2 \\ 0 \end{bmatrix} = x_2 \begin{bmatrix} 1 \\ 1 \\ 0 \end{bmatrix}$.

Therefore $\lambda = 3$ has algebraic and geometric multiplicity 1. The matrix is not defective.

15. The given matrix A has characteristic equation $0 = \det(A - tI) = -(t-2)^2(t-1)$ so the eigenvalues are $\lambda = 1$ and $\lambda = 2$. The system of equations $(A - I)\mathbf{x} = \theta$ has solution $x_1 = -3x_3, x_2 = -x_3, x_3$arbitrary so the eigenvectors for $\lambda = 1$ are the nonzero vectors of the form $\mathbf{x} = [-3x_3, -x_3, x_3]^T$.For the system $(A - 2I)\mathbf{x} = \theta$, x_1 and x_2 are arbitrary and $x_3 = 0$. The eigenvectors for $\lambda = 2$ are the nonzero vectors of the form $\mathbf{x}$ $= x_1 \begin{bmatrix} 1 \\ 0 \\ 0 \end{bmatrix} + x_2 \begin{bmatrix} 0 \\ 1 \\ 0 \end{bmatrix}$. The matrix is not defective.

17. The given matrix A has characteristic polynomial $p(t) =$ $-(t+1)(t-1)(t-2)$ so the eigenvalues for A are $\lambda = -1, \lambda = 1, \lambda = 2$. The system of equations $(A+I)\mathbf{x} = \theta$ has solution $x_1 = (1/2)x_3, x_2 = x_3, x_3$ arbitrary so the eigenvectors for $\lambda = -1$ are the nonzero vectors of the form $\mathbf{x} = [(1/2)x_3, x_3, x_3]^T$. The system of equations $(A - I)\mathbf{x} = \theta$ has solution $x_1 = -3x_2, x_2$ arbitrary,$x_3 = -7x_2$ so the eigenvectors for $\lambda = 1$ are the nonzero vectors of the form $\mathbf{x} = [-3x_2, x_2, -7x_2]^T$. The system of equations $(A - 2I)\mathbf{x} = \theta$ has solution $x_1 = (1/2)x_3, x_2 = (-1/2)x_3$ so the eigenvectors for $\lambda = 2$ are the nonzero vectors of the form $\mathbf{x} = [(1/2)x_3, (-1/2)x_3, x_3]^T$. The matrix is not defective.

19. The characteristic polynomial for $A$is $p(t) = -(t-1)^2(t-2)$ so the eigenvalues for A are $\lambda = 1$ and $\lambda = 2$. The vectors $\mathbf{u_1} = [1,0,0]^T$ and $\mathbf{u_2} = [0,1,2]^T$are the eigenvectors for $\lambda = 1$ and $\mathbf{u_3} = [1,2,3]^T$ is an eigenvector for $\lambda = 2$. Moreover $\mathbf{x} = \mathbf{u_1} + 2\mathbf{u_2} + \mathbf{u_3}$ so $A^{10}\mathbf{x} = (1)^{10}\mathbf{u_1} + 2(1)^{10}\mathbf{u_2} + (2)^{10}\mathbf{u_3} = [1025, 2050, 3076]^T$.

21. $P = P^{-1}P^2 = P^{-1}P = I$.

23. $P^2 = (\mathbf{u}\mathbf{u}^T)(\mathbf{u}\mathbf{u}^T) = \mathbf{u}(\mathbf{u}^T\mathbf{u})\mathbf{u}^T = \mathbf{u}\mathbf{u}^T = P$.

25. $P^2 = (\mathbf{u}\mathbf{u}^{\mathrm{T}} + \mathbf{v}\mathbf{v}^{\mathrm{T}})(\mathbf{u}\mathbf{u}^{\mathrm{T}} + \mathbf{v}\mathbf{v}^{\mathrm{T}}) = \mathbf{u}\,(\mathbf{u}^{\mathrm{T}}\mathbf{u})\mathbf{u}^{\mathrm{T}} + \mathbf{u}\,(\mathbf{u}^{\mathrm{T}}\mathbf{v})\mathbf{v}^{\mathrm{T}} +$
$\mathbf{v}\,(\mathbf{v}^{\mathrm{T}}\mathbf{u})\mathbf{u}^{\mathrm{T}} + \mathbf{v}\,(\mathbf{v}^{\mathrm{T}}\mathbf{v})\mathbf{v}^{\mathrm{T}} = \mathbf{u}\mathbf{u}^{\mathrm{T}} + \mathbf{v}\mathbf{v}^{\mathrm{T}} = P.$

27. (a) A has eigenvalues $\lambda = 1$ and $\lambda = 3$ with corresponding eigenvectors $\mathbf{u_1} = [1/\sqrt{2}, 1/\sqrt{2}]^{\mathrm{T}}$ and $\mathbf{u_2} = [-1/\sqrt{2}, 1/\sqrt{2}]^{\mathrm{T}}$, respectively, where $\| \mathbf{u_1} \| = \| \mathbf{u_2} \| = 1$. It is easily checked that $\mathbf{u_1}\mathbf{u_1}^{\mathrm{T}} + 3\mathbf{u_2}\mathbf{u_2}^{\mathrm{T}} = A$.

    (b) A has eigenvalues $\lambda = -1$ånd $\lambda = 3$ with corresponding eigenvectors $\mathbf{u_1} = [-1/\sqrt{2}, 1/\sqrt{2}]^{\mathrm{T}}$ and $\mathbf{u_2} = [1/\sqrt{2}, 1/\sqrt{2}]^{\mathrm{T}}$, respectively, where $\| \mathbf{u_1} \| = \| \mathbf{u_2} \| = 1$.İt is easily checked that $-\mathbf{u_1}\mathbf{u_1}^{\mathrm{T}} + 3\mathbf{u_2}\mathbf{u_2}^{\mathrm{T}} = A$.

    (c) A has eigenvalues $\lambda = 4$ and $\lambda = -1$ with corresponding eigenvectors $\mathbf{u_1} = [2/\sqrt{5}, 1/\sqrt{5}]^{\mathrm{T}}$ and $\mathbf{u_2} = [1/\sqrt{5}, -2/\sqrt{5}]^{\mathrm{T}}$, respectively, where $\| \mathbf{u_1} \| = \| \mathbf{u_2} \| = 1$.İt is easily checked that $4\mathbf{u_1}\mathbf{u_1}^{\mathrm{T}} - \mathbf{u_2}\mathbf{u_2}^{\mathrm{T}} = A$.

3.6 Complex Eigenvalues and Eigenvectors

1. $\overline{u} = 3 + 2i$.

3. $u + \overline{v} = 7 - 3i$.

5. $u + \overline{u} = 6$.

7. $v\overline{v} = 17$.

9. $s^2 - w = -5 + 5i$.

11. $\overline{u}w^2 = 17 - 6i$.

13. $u/v = (u\overline{v})/(v\overline{v}) = 10/17 - (11/17)i$.

15. $s/z = (s\overline{z})/(z\overline{z}) = 3/2 + (1/2)i$.

17. $w + iz = 1$.

19. For the given matrix A the characteristic polynomial is $p(t) = t^2 - 8t + 20$ and the eigenvalues are $\lambda = 4 + 2i$ and $\overline{\lambda} = 4 - 2i$. The system of equations $(A - (4 + 2i)I)\mathbf{x} = \theta$ is given by

$$\begin{array}{rcrcl} (2 - 2i)x_1 & + & 8x_2 & = & 0 \\ -x_1 & + & (-2 - 2i)x_2 & = & 0 \end{array}.$$

The solution is $x_1 = (-2 - 2i)x_2, x_2$årbitrary. Thus the eigenvectors for $\lambda = 4 + 2i$åre the nonzero vectors of the form
$\mathbf{x} = [(-2 - 2i)x_2, x_2]^{\mathrm{T}}$. By Theorem 16, $\overline{\mathbf{x}}$ is an eigenvector corresponding to $\overline{\lambda}$.

21. The given matrix A has characteristic polynomial $p(t) = t^2 + 1$ so the eigenvalues are $\lambda = i$ and $\overline{\lambda} = -i$. The system $(A - iI)\mathbf{x} = \theta$ is given by

$$\begin{array}{rcrcl} (-2-i)x_1 & - & x_2 & = & 0 \\ 5x_1 & + & (2-i)x_2 & = & 0 \end{array}.$$

The solution is $x_1 = ((-2+i)/5)x_2$, so the eigenvectors for $\lambda = i$ are the nonzero vectors of the form $\mathbf{x} = [((-2+i)/5)x_2, x_2]^{\mathrm{T}}$.

By Theorem 16, $\overline{\mathbf{x}}$ is an eigenvector for $\overline{\lambda} = -i$.

23. The given matrix A has characteristic polynomial $p(t) = -(t-2)(t^2 - 4t + 13)$ so the eigenvalues for A are $\lambda_1 = 2, \lambda_2 = 2 + 3i$, and $\overline{\lambda}_2 = 2 - 3i$. The system $(A - 2I)\mathbf{x} = \theta$ is given by

$$\begin{array}{rcrcrcl} -x_1 & - & 4x_2 & - & x_3 & = & 0 \\ 3x_1 & & & + & 3x_3 & = & 0 \\ x_1 & + & x_2 & + & x_3 & = & 0 \end{array}.$$

The solution is $x_1 = -x_3, x_2 = 0, x_3$ arbitrary so the eigenvectors corresponding to $\lambda_1 = 2$ are of the form $\mathbf{x} = [-x_3, 0, x_3]^{\mathrm{T}}$. The system $[A - (2+3i)I]\mathbf{x} = \theta$ is given by

$$\begin{array}{rcrcrcl} (-1-3i)x_1 & - & 4x_2 & - & x_3 & = & 0 \\ 3x_1 & - & 3ix_2 & + & 3x_3 & = & 0 \\ x_1 & + & x_2 & + & (1-3i)x_3 & = & 0 \end{array}.$$

The solution is $x_1 = (-5/2 + (3/2)i)x_3$ and $x_2 = (3/2 + (3/2)i)x_3$ so the eigenvectors for $\lambda_2 = 2 + 3i$ are the nonzero vectors of the form $\mathbf{x} = [(-5/2 + (3/2)i)x_3, (3/2 + (3/2)i)x_3, x_3]^{\mathrm{T}}$. By Theorem 16, $\overline{\mathbf{x}}$ is an eigenvector for $\overline{\lambda}_2 = 2 - 3i$.

25. $x = 2 - i, y = 3 - 2i$.

27. $\overline{\mathbf{x}}^{\mathrm{T}}\mathbf{x} = (1-i)(1+i) + 2(2) = 6$ so $\|\mathbf{x}\| = \sqrt{6}$.

29. $\overline{\mathbf{x}}^{\mathrm{T}}\mathbf{x} = (1+2i)(1-2i) + (-i)(i) + (3-i)(3+i) = 5 + 1 + 10 = 16$. Thus $\|\mathbf{x}\| = \sqrt{16} = 4$.

31. $\lambda_1 = -1.4937 + 1.2616i$, $\mathbf{x}_1 = \begin{bmatrix} 0.5835 - 0.1460i \\ 0.1650 - 0.4762i \\ -0.4369 + 0.4397i \end{bmatrix}$; $\lambda_2 = -1.4937 - 1.2616i$, $\mathbf{x}_2 = \begin{bmatrix} 0.5835 + 0.1460i \\ 0.1650 + 0.4762i \\ -0.4369 - 0.4397i \end{bmatrix}$; $\lambda_3 = 10.9873$, $\mathbf{x}_3 = \begin{bmatrix} -0.4486 \\ -0.7312 \\ -0.5139 \end{bmatrix}$. In each case, the eigenvectors are chosen to have length 1.

33. $\lambda_1 = 1.1857 + 2.6885i$, $\mathbf{x}_1 = \begin{bmatrix} -0.0781 - 0.6033i \\ -0.3495 + 0.5754i \\ 0.1199 - 0.1125i \\ 0.1963 + 0.3334i \end{bmatrix}$;

$\lambda_2 = 1.1857 - 2.6885i$, $\mathbf{x}_2 = \begin{bmatrix} -0.0781 + 0.6033i \\ -0.3495 - 0.5754i \\ 0.1199 + 0.1125i \\ 0.1963 - 0.3334i \end{bmatrix}$;

$\lambda_3 = 16.8037$, $\mathbf{x}_3 = \begin{bmatrix} -0.5484 \\ -0.0550 \\ -0.1746 \\ -0.8160 \end{bmatrix}$;

$\lambda_4 = 4.8249$, $\mathbf{x}_4 = \begin{bmatrix} 0.7046 \\ -0.6728 \\ -0.2027 \\ -0.0995 \end{bmatrix}$. For $i = 1, 2, 3, 4$, $\|\mathbf{x_i}\| = 1$.

35. Let $z = a + bi$ and $w = c + di$.

(a) $z + w = (a + c) + (b + d)i$ so $\overline{z + w} = (a + c) - (b + d)i = \overline{z} + \overline{w}$.

(b) $zw = (ac - bd) + (ad + bc)i$ so $\overline{zw} = (ac - bd) - (ad + bc)i$. Therefore $\overline{z}\,\overline{w} = (a - bi)(c - di) = (ac - bd) - (ad + bc)i = \overline{zw}$.

(c) $z + \overline{z} = (a + bi) + (a - bi) = 2a$.

(d) $z - \overline{z} = (a + bi) - (a - bi) = 2bi$.

(e) $z\overline{z} = (a + bi)(a - bi) = a^2 - b^2 i^2 = a^2 + b^2$.

3.7 Similarity Transformations and Diagonalization

1. A has eigenvalues $\lambda = 1$ and $\lambda = 3$ with corresponding eigenvectors $\mathbf{u_1} = [1, 1]^{\mathrm{T}}$ and $\mathbf{u_2} = [-1, 1]^{\mathrm{T}}$, respectively. If $S = [\mathbf{u_1}, \mathbf{u_2}]$ then $S^{-1}AS = D$ where $D = \begin{bmatrix} 1 & 0 \\ 0 & 3 \end{bmatrix}$. Now $D^5 = \begin{bmatrix} 1 & 0 \\ 0 & 243 \end{bmatrix}$ so $A^5 = SD^5S^{-1} = \begin{bmatrix} 122 & -121 \\ -121 & 122 \end{bmatrix}$.

3. A has only one eigenvalue, $\lambda = -1$. The corresponding eigenvectors are the nonzero vectors of the form $\mathbf{x} = [x_2, x_2]^{\mathrm{T}}$. Since we cannot find a linearly independent set $\{\mathbf{u_1}, \mathbf{u_2}\}$ of eigenvectors for A, A is not diagonalizable.

5. A has eigenvalues $\lambda = 1$ and $\lambda = 2$ with corresponding eigenvectors $\mathbf{u_1} = [1, -10]^T$ and $\mathbf{u_2} = [0, 1]^T$, respectively. If $S = [\mathbf{u_1}, \mathbf{u_2}]$ then $S^{-1}AS = D$ where $D = \begin{bmatrix} 1 & 0 \\ 0 & 2 \end{bmatrix}$. Thus

$$A^5 = SD^5S^{-1} = S\begin{bmatrix} 1 & 0 \\ 0 & 32 \end{bmatrix}S^{-1} = \begin{bmatrix} 1 & 0 \\ 310 & 32 \end{bmatrix}.$$

7. A has eigenvalue $\lambda = 1$ with algebraic multiplicity 3. The eigenvectors for $\lambda = 1$ have the form $\mathbf{x} = [x_2 + 2x_3, x_2, x_3]^T$. In particular we cannot obtain 3 linearly independent eigenvectors so A is not diagonalizable.

9. A has eigenvalues $1, 2,$ and -1 with corresponding eigenvectors $\mathbf{u_1} = [-3, 1, -7]^T, \mathbf{u_2} = [-1, 1, -2]^T, \mathbf{u_3} = [1, 2, 2]^T$, respectively. If $S = [\mathbf{u_1}, \mathbf{u_2}, \mathbf{u_3}]$ then

$$S^{-1} = \begin{bmatrix} 2 & 0 & -1 \\ 16/3 & 1/3 & 7/3 \\ 5/3 & 1/3 & -2/3 \end{bmatrix} \text{ and } S^{-1}AS = D, \text{where } D = \begin{bmatrix} 1 & 0 & 0 \\ 0 & 2 & 0 \\ 0 & 0 & -1 \end{bmatrix}.$$

$$A^5 = SD^5S^{-1} = \begin{bmatrix} 163 & -11 & -71 \\ -172 & 10 & 75 \\ 324 & -22 & -141 \end{bmatrix}.$$

11. A has eigenvalue $\lambda = 1$ with algebraic multiplicity 2 and geometric multiplicity 1 so A is not diagonalizable.

13. $\mathbf{q_1}^T\mathbf{q_2} = 0$ and $\mathbf{q_1}^T\mathbf{q_1} = \mathbf{q_2}^T\mathbf{q_2} = 1$ so Q is orthogonal.

15. $\mathbf{q_1}^T\mathbf{q_1} = 5$ so Q is not orthogonal.

17. $0 = \mathbf{q_1}^T\mathbf{q_2} = \mathbf{q_1}^T\mathbf{q_3} = \mathbf{q_2}^T\mathbf{q_3}$ and $1 = \mathbf{q_1}^T\mathbf{q_1} = \mathbf{q_2}^T\mathbf{q_2} = \mathbf{q_3}^T\mathbf{q_3}$ so Q is orthogonal.

19. If Q is orthogonal then $2\alpha^2 = 1, 6\beta^2 = 1, a^2+b^2+c^2 = 1, \alpha a+\alpha c = 0,$ and $\beta a+2\beta b-\beta c = 0$. This implies that $\alpha = 1/\sqrt{2}, \beta = 1/\sqrt{6}, a = -c$ and $b = c$, where $c = \pm 1/\sqrt{3}$. Thus we one choice for Q is

$$Q = \begin{bmatrix} 1/\sqrt{2} & 1/\sqrt{6} & -1/\sqrt{3} \\ 0 & 2/\sqrt{6} & 1/\sqrt{3} \\ 1/\sqrt{2} & -1/\sqrt{6} & 1/\sqrt{3} \end{bmatrix}.$$

21. $$Q = \begin{bmatrix} -0.8807 & 0.4332 & 0.1918 \\ 0.1849 & 0.6870 & -.7028 \\ 0.4361 & 0.5835 & 0.6851 \end{bmatrix} \quad T = \begin{bmatrix} 0.5048 & 2.9498 & -1.4966 \\ 0.0 & 8.3443 & 0.2429 \\ 0.0 & 0.0 & -2.8491 \end{bmatrix}$$

23. $$Q = \begin{bmatrix} -0.5276 & -0.5463 & -0.6406 & 0.1130 \\ -0.4235 & -0.4440 & 0.6477 & -0.4516 \\ -0.3315 & 0.0292 & 0.3989 & 0.8545 \\ -0.6576 & 0.7096 & -0.1044 & -0.2306 \end{bmatrix}$$

$$T = \begin{bmatrix} 19.2422 & 1.4541 & -3.5019 & 0.0983 \\ 0.0 & -3.8383 & -3.4646 & -0.5055 \\ 0.0 & 0.0 & 4.8409 & 3.5148 \\ 0.0 & 0.0 & 0.0 & 0.7552 \end{bmatrix}$$

25. (a) $(S^{-1}AS)^2 = (S^{-1}AS)(S^{-1}AS) = S^{-1}A(SS^{-1})AS = S^{-1}A^2S.$
$(S^{-1}AS)^3 = (S^{-1}AS)^2(S^{-1}AS) = (S^{-1}A^2S)(S^{-1}AS) = S^{-1}A^2(SS^{-1})AS = S^{-1}A^3S.$
(b) Suppose $(S^{-1}AS)^n = S^{-1}A^nS$ for some integer $n \geq 1$. Then $(S^{-1}AS)^{n+1} = (S^{-1}AS)^n(S^{-1}AS) = (S^{-1}A^nS)(S^{-1}AS) = S^{-1}A^n(SS^{-1})AS = S^{-1}A^{n+1}S$. By induction $(S^{-1}AS)^k = S^{-1}A^kS$ for any positive integer k.

31. $\mathbf{y} = [b/\sqrt{a^2+b^2}, -a/\sqrt{a^2+b^2}]^T$ is one choice. $-\mathbf{y}$ is another choice.

33. If $\mathbf{u} = [1/\sqrt{2}, -1/\sqrt{2}]^T$ then $A\mathbf{u} = 2\mathbf{u}$ and $\mathbf{u}^T\mathbf{u} = 1$. If $\mathbf{v} = [1/\sqrt{2}, 1/\sqrt{2}]^T$ then $\mathbf{u}^T\mathbf{v} = 0$ and $\mathbf{v}^T\mathbf{v} = 1$. If $Q = [\mathbf{u}, \mathbf{v}]$ then $Q^TAQ = \begin{bmatrix} 2 & -2 \\ 0 & 2 \end{bmatrix}$.

35. If $\mathbf{u} = [1/\sqrt{2}, 1/\sqrt{2}]^T$ then $A\mathbf{u} = \mathbf{u}$ and $\mathbf{u}^T\mathbf{u} = 1$. If $\mathbf{v} = [1/\sqrt{2}, -1/\sqrt{2}]^T$ then $\mathbf{u}^T\mathbf{v} = 0$ and $\mathbf{v}^T\mathbf{v} = 1$. If $Q = [\mathbf{u}, \mathbf{v}]$ then
$Q^TAQ = \begin{bmatrix} 1 & 0 \\ 0 & 3 \end{bmatrix}$.

3.8 Applications

1. $\mathbf{x_1} = \begin{bmatrix} 4 \\ 2 \end{bmatrix}, \mathbf{x_2} = \begin{bmatrix} 2 \\ 4 \end{bmatrix}, \mathbf{x_3} = \begin{bmatrix} 4 \\ 2 \end{bmatrix}, \mathbf{x_4} = \begin{bmatrix} 2 \\ 4 \end{bmatrix}.$

5. $\mathbf{x_1} = \begin{bmatrix} 7 \\ 1 \end{bmatrix}, \mathbf{x_2} = \begin{bmatrix} 11 \\ 8 \end{bmatrix}, \mathbf{x_3} = \begin{bmatrix} 43 \\ 19 \end{bmatrix}, \mathbf{x_4} = \begin{bmatrix} 119 \\ 62 \end{bmatrix}.$

7. A has eigenvalues $\lambda_1 = 1$ and $\lambda_2 = -1$ with corresponding eigenvectors $\mathbf{u_1} = [1,1]^T$ and $\mathbf{u_2} = [-1,1]^T$, respectively. $\mathbf{x_0} = 3\mathbf{u_1} + \mathbf{u_2}$ so $\mathbf{x_k} = 3(1)^k\mathbf{u_1} + (-1)^k\mathbf{u_2} = [3 + (-1)^{k+1}, 3 + (-1)^k]^T$. In particular $\mathbf{x_4} = [2,4]^T = \mathbf{x_{10}}$. The sequence $\{\mathbf{x_k}\}$ has no limit but $\| \mathbf{x_k} \| = \sqrt{20}$ for all k.

9. A has eigenvalues $\lambda_1 = 1$ and $\lambda_2 = 0.25$ with corresponding eigenvectors $\mathbf{u_1} = [1,2]^T$ and $\mathbf{u_2} = [-1,1]^T$, respectively. $\mathbf{x_0} = 64\mathbf{u_1} - 64\mathbf{u_2}$ so $\mathbf{x_k} = 64(1)^k\mathbf{u_1} - 64(0.25)^k\mathbf{u_2} = [64 + 64(0.25)^k, 128 - 64(0.25)^k]^T$. In particular $\mathbf{x_4} = [64.25, 127.75]^T$ and $\mathbf{x_{10}} = [64.00006, 127.99994]^T$. The sequence $\{\mathbf{x_k}\}$ converges to $\mathbf{x}^* = [64, 128]^T$.

11. A has eigenvalues $\lambda_1 = 3$ and $\lambda_2 = -1$ with corresponding eigenvectors $\mathbf{u_1} = [2,1]^{\mathrm{T}}$ and $\mathbf{u_2} = [-2,1]^{\mathrm{T}}$, respectively. $\mathbf{x_0} =$ $(3/4)\mathbf{u_1} + (5/4)\mathbf{u_2}$ so $\mathbf{x_k} = (3/4)(3)^k\mathbf{u_1} + (5/4)(-1)^k\mathbf{u_2} =$ $(1/4)[2(3)^{k+1} - 10(-1)^k, 3^{k+1} + 5(-1)^k]^{\mathrm{T}}$. In particular $\mathbf{x_4} =$ $[119, 62]^{\mathrm{T}}$ and $\mathbf{x_{10}} = [88571, 44288]^{\mathrm{T}}$. The sequence $\{\mathbf{x_k}\}$ has no limit and $\lim_{k\to\infty} \|\mathbf{x_k}\| = \infty$.

13. A has eigenvalues $\lambda_1 = 2, \lambda_2 = 1, \lambda_3 = -1$ with corresponding eigenvectors $\mathbf{u_1} = [1,-1,2]^{\mathrm{T}}, \mathbf{u_2} = [3,-1,7]^{\mathrm{T}}$, and $\mathbf{u_3} = [1,2,2]^{\mathrm{T}}$, respectively. $\mathbf{x_0} = 2\mathbf{u_1} + 2\mathbf{u_2} - 5\mathbf{u_3}$, so $\mathbf{x_k} = 2^{k+1}\mathbf{u_1} + 2\mathbf{u_2} + 5(-1)^{k+1}\mathbf{u_3}$;

$$\text{thus}: \mathbf{x_k} = \begin{bmatrix} 2^{k+1} + 6 + 5(-1)^{k+1} \\ -2^{k+1} - 2 + 10(-1)^{k+1} \\ 2^{k+2} + 14 + 10(-1)^{k+1} \end{bmatrix}.$$

In particular $\mathbf{x_4} = [33, -44, 68]^{\mathrm{T}}$ and $\mathbf{x_{10}} = [2049, -2060, 4100]^{\mathrm{T}}$. The sequence $\{\mathbf{x_k}\}$ has no limit and $\lim_{k\to\infty} \|\mathbf{x_k}\| = \infty$.

15. If $\mathbf{x}(t) = \begin{bmatrix} u(t) \\ v(t) \end{bmatrix}$ then $\mathbf{x}'(t) = A\mathbf{x}(t)$ for $A = \begin{bmatrix} 5 & -6 \\ 3 & -4 \end{bmatrix}$. Eigenvalues and corresponding eigenvectors for A are $\lambda_1 = 2$, $\mathbf{u}_1 = \begin{bmatrix} 2 \\ 1 \end{bmatrix}$ and $\lambda_2 = -1$, $\mathbf{u}_2 = \begin{bmatrix} 1 \\ 1 \end{bmatrix}$. Setting $\mathbf{x}(t) =$ $ae^{2t}\mathbf{u}_1 + be^{-t}\mathbf{u}_2$ and $\mathbf{x}(0) = \begin{bmatrix} 4 \\ 1 \end{bmatrix}$ yields $a = 3$ and $b = -2$. Thus. $\mathbf{x}(t) = 3e^{2t}\mathbf{u}_1 - 2e^{-t}\mathbf{u}_2$. Equivalently, $u(t) = 6e^{2t} - 2e^{-t}$ and $v(t) = 3e^{2t} - 2e^{-t}$.

17. The matrix $A = \begin{bmatrix} 1 & 1 & 1 \\ 0 & 3 & 3 \\ -2 & 1 & 1 \end{bmatrix}$ has eigenvalues $\lambda_1 = 0$, $\lambda_2 = 2$, $\lambda_3 = 3$. Corresponding eigenvectors are $\mathbf{u}_1 = [0,-1,1]^T$, $\mathbf{u}_2 = [-2,-3,1]^T$, $\mathbf{u}_3 = [1,2,0]^T$, respectively. The solution is $\mathbf{x}(t) = 2\mathbf{u}_1 - e^{2t}\mathbf{u}_2 + e^{3t}\mathbf{u}_3$. Equivalently, $u(t) = 2e^{2t} + e^{3t}$, $v(t) = -2 + 3e^{2t} + 2e^{3t}$, and $w(t) = 2 - e^{2t}$.

19. (a) The eigenvectors corresponding to $\lambda = 1$ have the form $\mathbf{x} = [a, 0], a \neq 0$. In particular $\lambda = 1$ has algebraic multiplicity 2 and geometric multiplicity 1. Thus A is defective.

    (b) For $k = 0$ we have $\mathbf{x_0} = [1,1]^{\mathrm{T}} = [2(0)+1, 1]^{\mathrm{T}}$. Suppose $\mathbf{x_m} = [2m+1, 1]^{\mathrm{T}}$ for some integer $m \geq 0$. Then $\mathbf{x_{m+1}} =$ $A\mathbf{x_m} = [2m+3, 1]^{\mathrm{T}} = [2(m+1)+1, 1]^{\mathrm{T}}$. It follows that $\mathbf{x_k} = [2k+1, 1]^{\mathrm{T}}$ for each $k \geq 0$.

21. For $\alpha = -0.18$ A has eigenvalues $\lambda_1 = 1$ and $\lambda_2 = -0.18$ with corresponding eigenvectors $\mathbf{u_1} = [3, 10]^{\mathrm{T}}$ and $\mathbf{u_2} = [10, -6]^{\mathrm{T}}$, respectively. Moreover $\mathbf{x_0} = (16/118)\mathbf{u_1} + (7/118)\mathbf{u_2}$. Therefore $\mathbf{x_k} = (16/118)\mathbf{u_1} + (7/118)(-0.18)^k\mathbf{u_2}$. It follows that $\lim_{x\to\infty} \mathbf{x_k} = (16/118)\mathbf{u_1}$

Chapter 4

Vector Spaces and Linear Transformations

4.1 Introduction (No exercises)

4.2 Vector Spaces

1. $\mathbf{u}-2\mathbf{v}=\begin{bmatrix} 0 & -7 & 5 \\ -11 & -3 & -12 \end{bmatrix}$; $\mathbf{u}-(2\mathbf{v}-3\mathbf{w})=\begin{bmatrix} 12 & -22 & 38 \\ -50 & -6 & -15 \end{bmatrix}$;

   $-2\mathbf{u}-\mathbf{v}+3\mathbf{w}=\begin{bmatrix} 7 & -21 & 28 \\ -42 & -7 & -14 \end{bmatrix}$.

3. $\mathbf{u}-2\mathbf{v}=e^x-2\sin x$; $\mathbf{u}-(2\mathbf{v}-3\mathbf{w})=e^x-2\sin x+3\sqrt{x^2+1}$; $-2\mathbf{u}-\mathbf{v}+3\mathbf{w}=-2e^x-\sin x+3\sqrt{x^2+1}$;

5. Note that $c_1\mathbf{u}+c_2\mathbf{v}+c_3\mathbf{w}=(c_1+c_2)x^2+(2c_2+2c_3)x+(-2c_1-c_2+c_3)$. Thus $c_1\mathbf{u}+c_2\mathbf{v}+c_3\mathbf{w}=x^2+6x+1$ if and only if

$$\begin{array}{rcrcrcl} c_1 & + & c_2 & & & = & 1 \\ & & 2c_2 & + & 2c_3 & = & 0 \\ -2c_1 & - & c_2 & + & c_3 & = & 0. \end{array}$$

   Solving yields $c_1=-2+c_3, c_2=3-c_3, c_3$ arbitrary. One choice is $c_1=-2, c_2=3, c_3=0$ and a direct calculation shows that $-2\mathbf{u}+3\mathbf{v}=x^2+6x+1$. Similarly $c_1\mathbf{u}+c_2\mathbf{v}+c_3\mathbf{w}=x^2$ if and only if

$$\begin{array}{rcrcrcl} c_1 & + & c_2 & & & = & 1 \\ & & 2c_2 & + & 2c_3 & = & 6 \\ -2c_1 & - & c_2 & + & c_3 & = & 1. \end{array}$$

The system is easily seen to be inconsistent.

7. S is not a vector space. None of properties (c1), (c2), (a3), and (a4) of Definition 1 is satisfied. For example $\mathbf{v} = [1,0,0,0]^{\mathrm{T}}$ and
$\mathbf{w} = [0,0,0,1]^{\mathrm{T}}$ are in S but $\mathbf{u}+\mathbf{w}$ is not in S.

9. P is not a vector space. Properties (c1), (c2), and (a3) of Definition 1 fail to hold in P. For example $p(x) = 1 + 2x^2$ and $q(x) = x - 2x^2$ are in P but $p(x) + q(x)$ is not in P.

11. P is not a vector space (cf. Exercise 9).

13. S is a vector space.

15. S is not a vector space. Properties (c1), (c2), and (a3) of Definition 1 fail to hold in S.

17. Let $A = \begin{bmatrix} 1 & 0 \\ 0 & 1 \end{bmatrix}$ and let $B = \begin{bmatrix} 0 & 1 \\ 1 & 0 \end{bmatrix}$. Then A and B are in Q but $A+B$ is not in Q. Also $0A$, the (2 x 2) zero matrix, is not in Q.

25. F is a vector space. Since F is a subset of $C[-1,1]$ and $C[-1,1]$ is a vector space, properties (a1), (a2), (m1), (m2), (m3), and (m4) hold in F. Now let $g(x), h(x)$ be in F; that is $g(x)$ and $h(x)$ are continuous, $g(-1) = g(1)$ and $h(-1) = h(1)$. It follows that $(g+h)(x) = g(x) + h(x)$ is continuous and $(g+h)(-1) = g(-1) + h(-1) = g(1) + h(1) = (g+h)(1)$. Therefore $(g+h)(x)$ is in F and property (c1) holds. If a is a scalar then $(ag)(x) = ag(x)$ is continuous and $(ag)(-1) = ag(-1) = ag(1) = (ag)(1)$, so $(ag)(x)$ is in F. This verifies that (c2) holds. The zero vector in $C[-1,1]$ is the function θ defined by $\theta(x) = 0$ for all x, $-1 \le x \le 1$. In particular $\theta(-1) = 0 = \theta(1)$ so $\theta(x)$ is in F. Thus $\theta(x)$ is also the zero vector for F and (a3) is satisfied. Property (a4) is an immediate consequence of (c2) since $-g(x) = (-1)g(x)$ for $g(x)$ in $C[-1,1]$. Since F satisfies the properties of Definition 1, F is a vector space.

27. F is not a vector space. For example set $f(x) = 2x - 1$ and $g(x) = 2x^2 - 1$. Then $f(x)$ and $g(x)$ are in F whereas $f(x) + g(x) = 2x^2 + 2x - 2$ is not.

29. F is a vector space. As in Exercise 25, it suffices to check that properties (c1), (c2), (a3), and (a4) of Definition 1 are satisfied. To check (c1) for example, let $f(x)$ and $g(x)$ be in F . Then $\int_{-1}^{1}[f(x) + g(x)]dx = \int_{-1}^{1} f(x)dx + \int_{-1}^{1} g(x)dx = 0 + 0 = 0$, so $f(x) + g(x)$ is in F .

4.3 Subspaces

1. W is not a subspace of V. None of the properties (s1), (s2), and (s3) of Theorem 2 is satisfied. For example, if $A = \begin{bmatrix} 1 & 0 & 0 \\ 0 & 0 & 0 \end{bmatrix}$ and $B = \begin{bmatrix} 0 & 0 & 1 \\ 0 & 0 & 0 \end{bmatrix}$ then A and B are in

W but $A+B$ is not in W.

3. W is a subspace of V. Clearly the (2 x 3) zero matrix is in W. Let $A=[a_{ij}]$ and $B=[b_{ij}]$ be in W. Therefore $a_{11}-a_{12}=0, a_{12}+a_{13}=0, a_{23}=0, b_{11}-b_{12}=0, b_{12}+b_{13}=0, b_{23}=0$. Now $A+B$ is the (2 x 3) matrix $A+B=[c_{ij}]$, where $c_{ij}=a_{ij}+b_{ij}$. Thus $c_{11}-c_{12}=(a_{11}+b_{11})-(a_{12}+b_{12})=(a_{11}-a_{12})+(b_{11}-b_{12})=0$.Similarly, $c_{12}+c_{13}=0$ and $c_{23}=0$. This shows that $A+B$ is in W. If k is a scalar then kA is the (2 x 3) matrix $kA=[d_{ij}]$ where $d_{ij}=ka_{ij}$. Consequently $d_{11}-d_{12}=ka_{11}-ka_{12}=k(a_{11}-a_{12})=k0=0$. Likewise $d_{12}+d_{13}=0$ and $d_{23}=0$. Therefore kA is in W. It follows from Theorem 2 that W is a subspace of V.

5. W is a subspace of $\mathcal{P}_2$. If $\theta(x)=0+0x+0x^2$ is the zero polynomial then clearly $\theta(0)+\theta(2)=0$, so $\theta(x)$ is in W. Suppose $g(x)$ and $h(x)$ are in W; that is $g(0)+g(2)=0$ and $h(0)+h(2)=0$. Then $(g+h)(0)=+(g+h)(2)=[g(0)+h(0)]+[g(2)+h(2)]=[g(0)+g(2)]+[h(0)+h(2)]=0+0=0$ and it follows that $g(x)+h(x)$ is in W. If c is a scalar then $(cg)(0)+(cg)(2)=c[g(0)+g(2)]=0$ and hence $(cg)(x)$ is in W. By Theorem 2, W is a subspace of $\mathcal{P}_2$.

7. W is not a subspace of $\mathcal{P}_2$. For example if $p(x)=x^2+x-2$ and $q(x)=x^2-9$ then $p(x)$ and $q(x)$ are in W but $p(x)+q(x)=2x^2+x-11$ is not in W. Note that properties (s1) and s(3) of Theorem 2 are satisfied.

9. F is a subspace of $C[-1,1]$. First recall that the zero of $C[-1,1]$ is the function θ defined by $\theta(x)=0$ for $-1\le x\le 1$. Since $\theta(-1)=0=-\theta(1), \theta(x)$ is in F. Now assume that $g(x)$ and $h(x)$ are in F. Thus $g(-1)=-g(1)$ and $h(-1)=-h(1)$. It follows that

$$(g+h)(-1)=g(-1)+h(-1)=-g(1)-h(1)=-(g+h)(1).$$

Therefore $(g+h)(x)$ is in F. If c is a scalar then $(cg)(-1)=$

$c(g(-1))=c(-g(1))=-(cg)(1)$ so $(cg)(x)$ is in F. by Theorem 2, F is a subspace of $C[-1,1]$.

11. F is not a subspace of $C[-1,1]$. None of the properties (s1), s(2), (s3) of Theorem 2 is satisfied. For example if $g(x)$ and $h(x)$ are in F then $(g+h)(-1)=g(-1)+h(-1)=-2+(-2)=-4$ so $(g+h)(x)$ is not in F .

13. F is a subspace of $C^2[-1,1]$. If $\theta(x)$ is the zero function then $\theta''(x)=\theta(x)=0$ for $-1\le x\le 1$. In particular $\theta''(0)=0$ so $\theta(x)$ is in F. Let $g(x)$ and $h(x)$ be in F; that is $g''(0)=0=h''(0)$. Therefore $(g+h)''(0)=g''(0)+h''(0)=0$, so $(g+h)(x)$ is in F . If c is a scalar then $(cg)''(0)=cg''(0)=0$ and $(cg)(x)$ is in F . By Theorem 2, F is a subspace of $C^2[-1,1]$.

15. F is not a subspace of $C^2[-1,1]$. None of the properties (s1), (s2), s(3) of Theorem 2 is satisfied. For example suppose $g(x)$ and $h(x)$ are in F . Then $g''(x)+g(x)=\sin x$ and

$h''(x)+h(x) = \sin x$ for $-1 \le x \le 1$. But $(g+h)''(x)+(g+h)(x) = 2\sin x$ for $-1 \le x \le 1$. Therefore $(g+h)(x)$ is not in F.

17. Note that $c_1p_1(x)+c_2p_2(x)+c_3p_3(x) = (c_1+2c_2+3c_3)+(2c_1+5c_2+8c_3)x+(c_1-2c_3)x^2$. Therefore $c_1p_1(x)+c_2p_2(x)+c_3p_3(x) = -1-3x+3x^2$ requires that $c_1+2c_2+3c_3 = -1, 2c_1+5c_2+8c_3 = -3$, and $c_1-2c_3 = 3$. Solving we obtain $c_1 = -1, c_2 = 3, c_3 = -2$ and it is easily verified that $p(x) = -p_1(x)+3p_2(x)-2p_3(x)$.

19. From the matrix equation $A = \sum_{i=1}^{4} c_iB_i$ we obtain the system of equations

$$\begin{array}{rcrcrcrcr} c_1 & + & c_2 & - & c_3 & + & c_4 & = & -2 \\ & & c_2 & - & 3c_3 & + & 2c_4 & = & -4 \\ 2c_1 & + & & & 4c_3 & - & c_4 & = & 1 \\ c_1 & + & 2c_2 & - & 4c_3 & + & c_4 & = & 0 \end{array}.$$

The solution is $c_1 = -1-2c_3, c_2 = 2+3c_3, c_3$ arbitrary, and $c_4 = -3$. Taking $c_3 = 0$ we see that $A = -B_1+2B_2-3B_4$.

21. $\cos 2x = (-1)\sin^2 x + (1)\cos^2 x$.

23. Let $p(x) = a_0+a_1x+a_2x^2+a_3x^3$ be in W. The constraints $p(1) = p(-1)$ and $p(2) = p(-2)$ imply that $a_0+a_1+a_2+a_3 = a_0-a_1+a_2-a_3$ and $a_0+2a_1+4a_2+8a_3 = a_0-2a_1+4a_2-8a_3$. This forces $a_1 = a_3 = 0$ while a_0 and a_2 are arbitrary. Thus $\{1, x^2\}$ is a spanning set for W.

25. For Exercise 2, a matrix $A = [a_{ij}]$ is in W if and only if A has the form $A = \begin{bmatrix} a_{12}-2a_{13} & a_{12} & a_{13} \\ a_{21} & a_{22} & a_{23} \end{bmatrix}$, where $a_{12}, a_{13}, a_{21}, a_{22}, a_{23}$ are arbitrary. Therefore

$$W = \mathrm{Sp}\left\{ \begin{bmatrix} 1 & 1 & 0 \\ 0 & 0 & 0 \end{bmatrix}, \begin{bmatrix} -2 & 0 & 1 \\ 0 & 0 & 0 \end{bmatrix}, \right.$$

$$\left. \begin{bmatrix} 0 & 0 & 0 \\ 1 & 0 & 0 \end{bmatrix}, \begin{bmatrix} 0 & 0 & 0 \\ 0 & 1 & 0 \end{bmatrix}, \begin{bmatrix} 0 & 0 & 0 \\ 0 & 0 & 1 \end{bmatrix} \right\}.$$

For Exercise 3, $A = [a_{ij}]$ is in W if and only if $A = \begin{bmatrix} a_{11} & a_{11} & -a_{11} \\ a_{21} & a_{22} & 0 \end{bmatrix}$, where a_{11}, a_{21}, and a_{22} are arbitrary. Therefore

$$W = \mathrm{Sp}\left\{ \begin{bmatrix} 1 & 1 & -1 \\ 0 & 0 & 0 \end{bmatrix}, \begin{bmatrix} 0 & 0 & 0 \\ 1 & 0 & 0 \end{bmatrix}, \begin{bmatrix} 0 & 0 & 0 \\ 0 & 1 & 0 \end{bmatrix} \right\}.$$

For Exercise 5 let $p(x) = a_0 + a_1x + a_2x^2$. The condition $p(0) + p(2) = 0$ implies that $2a_0 + 2a_1 + 4a_2 = 0$. Therefore $p(x) = (-a_1 - 2a_2) + a_1x + a_2x^2$ where a_1 and a_2 are arbitrary. It follows that $W = \text{Sp}\{-1 + \text{x}, -2 + \text{x}^2\}$.

For Exercise 6, $W = \text{Sp}\{\ 1, -4x + x^2\}$.

For Exercise 8, $W = \text{Sp}\{\ x, 1 - x^2\}$.

27. It is straightforward to show that $tr(A + B) = tr(A) + tr(B)$ and $tr(cA) = ctr(A)$. It then follows easily from Theorem 2 that W is a subspace of V. If A is in W then A has the form $A =$

$\begin{bmatrix} -a_{22} - a_{33} & a_{12} & a_{13} \\ a_{21} & a_{22} & a_{23} \\ a_{31} & a_{32} & a_{33} \end{bmatrix}$. It follows that $W = \text{Sp}\{\text{B}_1, \text{B}_2, \text{E}_{12}, \text{E}_{13}, \text{E}_{21}, \text{E}_{23}, \text{E}_{31}, \text{E}_{32}\}$

where $B_1 = \begin{bmatrix} -1 & 0 & 0 \\ 0 & 1 & 0 \\ 0 & 0 & 0 \end{bmatrix}$ and $B_2 = \begin{bmatrix} -1 & 0 & 0 \\ 0 & 0 & 0 \\ 0 & 0 & 1 \end{bmatrix}$.

29. For any (n x n) matrix $A, A = B + C$ where B and C are the matrices given in Exercise 28.

31. (a) A is in W if and only if A has the form $A = \begin{bmatrix} 0 & a_{12} & a_{13} \\ 0 & 0 & a_{23} \\ 0 & 0 & 0 \end{bmatrix}$, where a_{12}, a_{13}, a_{23} are arbitrary. Thus $W = \text{Sp}\{\text{E}_{12}, \text{E}_{13}, \text{E}_{23}\}$.

(b) A is in W if and only if A has form $A =$

$\begin{bmatrix} -a_{22} - a_{33} & -a_{23} & a_{13} \\ 0 & a_{22} & a_{23} \\ 0 & 0 & a_{33} \end{bmatrix}$. Therefore

$$W = \text{Sp}\left\{ \begin{bmatrix} -1 & 0 & 0 \\ 0 & 1 & 0 \\ 0 & 0 & 0 \end{bmatrix}, \begin{bmatrix} -1 & 0 & 0 \\ 0 & 0 & 0 \\ 0 & 0 & 1 \end{bmatrix}, \begin{bmatrix} 0 & -1 & 0 \\ 0 & 0 & 1 \\ 0 & 0 & 0 \end{bmatrix}, \begin{bmatrix} 0 & 0 & 1 \\ 0 & 0 & 0 \\ 0 & 0 & 0 \end{bmatrix} \right\}.$$

(c) A is in W if and only if A has the form $A = \begin{bmatrix} a_{11} & a_{11} & a_{13} \\ 0 & a_{22} & a_{13} \\ 0 & 0 & a_{22} \end{bmatrix}$.. Therefore

$$W = \text{Sp}\left\{ \begin{bmatrix} 1 & 1 & 0 \\ 0 & 0 & 0 \\ 0 & 0 & 0 \end{bmatrix}, \begin{bmatrix} 0 & 0 & 1 \\ 0 & 0 & 1 \\ 0 & 0 & 0 \end{bmatrix}, \right.$$

$$\left.\begin{bmatrix} 0 & 0 & 0 \\ 0 & 1 & 0 \\ 0 & 0 & 1 \end{bmatrix}\right\}.$$

(d) A is in W if and only if A has the form $A =$
$\begin{bmatrix} a_{11} & -a_{23} & a_{13} \\ 0 & a_{11} & a_{23} \\ 0 & 0 & a_{11} \end{bmatrix}$. Therefore $W = \text{Sp}\left\{\begin{bmatrix} 1 & 0 & 0 \\ 0 & 1 & 0 \\ 0 & 0 & 1 \end{bmatrix},\right.$
$\left.\begin{bmatrix} 0 & -1 & 0 \\ 0 & 0 & 1 \\ 0 & 0 & 0 \end{bmatrix}, \begin{bmatrix} 0 & 0 & 1 \\ 0 & 0 & 0 \\ 0 & 0 & 0 \end{bmatrix}\right\}.$

33. The equation $A = \sum_{i=1}^{4} x_i B_i$ implies that

$$\begin{array}{rcrcrcrcl} x_1 & + & 2x_2 & - & x_3 & + & x_4 & = & a \\ & & x_2 & + & 3x_3 & + & x_4 & = & b \\ x_1 & + & x_2 & - & 3x_3 & - & 2x_4 & = & c \\ -2x_1 & - & 2x_2 & + & 6x_3 & + & 5x_4 & = & d \end{array}.$$

Solving we obtain $x_1 = -6a+5b+37c+15d, x_2 = 3a-2b-17c-7d, x_3 = -a+b+5c+2d, x_4 = 2c+d$. Therefore $C = -12B_1 + 6B_2 - B_3 - B_4$ and $D = 8B_1 - 3B_2 + B_3 + B_4$.

4.4 Linear Independence, Bases, and Coordinates

1. If A is in W then $A = \begin{bmatrix} -b-c-d & b \\ c & d \end{bmatrix} = b\begin{bmatrix} -1 & 1 \\ 0 & 0 \end{bmatrix} +$
$c\begin{bmatrix} -1 & 0 \\ 1 & 0 \end{bmatrix} + d\begin{bmatrix} -1 & 0 \\ 0 & 1 \end{bmatrix}$. The set $\left\{\begin{bmatrix} -1 & 1 \\ 0 & 0 \end{bmatrix},\right.$
$\left.\begin{bmatrix} -1 & 0 \\ 1 & 0 \end{bmatrix}, \begin{bmatrix} -1 & 0 \\ 0 & 1 \end{bmatrix}\right\}$ is a basis for W.

5. For $p(x)$ in $W, p(x) = a_0 + a_1 x + (a_0 - 2a_1)x^2 = a_0(1 + x^2) +$
$a_1(x - 2x^2)$. The set $\{1 + x^2, x - 2x^2\}$ is a basis for W.

7. The set $\{x, x^2\}$ is a basis for W.

9. Let $p(x) = \sum_{i=0}^{4} a_i x^i$ be in V. The given constraints are as follows:

$$\begin{array}{lr} p(0) = 0: & a_0 = 0 \\ p'(1) = 0: & a_1 + 2a_2 + 3a_3 + 4a_4 = 0 \\ p''(-1) = 0: & 2a_2 - 6a_3 + 12a_4 = 0 \end{array}.$$

Solving yields $a_0 = 0, a_1 = -9a_3 + 8a_4, a_2 = 3a_3 - 6a_4, a_3$ and a_4 arbitrary. Thus $\{-9x + 3x^2 + x^3, 8x - 6x^2 + x^4\}$ is a basis for V.

11. If $A = [a_{ij}]$ is a (2 x 2) matrix then $A = a_{11}E_{11} + a_{12}E_{12} + a_{21}E_{21} + a_{22}E_{22}$ so B spans V. It is easy to see that B is a linearly independent set, so B is a basis for W.

13. (a) $[2, -1, 3, 2]^{\mathrm{T}}$ (b) $[1, 0, -1, 1]^{\mathrm{T}}$ (c) $[2, 3, 0, 0]^{\mathrm{T}}$.

15. The given matrices have coordinate vectors $\mathbf{u_1} = [2, 1, 2, 1]^{\mathrm{T}}$,
$\mathbf{u_2} = [3, 0, 0, 2]^{\mathrm{T}}, \mathbf{u_3} = [1, 1, 2, 1]^{\mathrm{T}}$, respectively. The equation $x_1\mathbf{u_1} + x_2\mathbf{u_2} + x_3\mathbf{u_3} = \theta$ has only the trivial solution so $\{\mathbf{u_1}, \mathbf{u_2}, \mathbf{u_3}\}$ is a linearly independent subset of R^4. By property (2) of Theorem 5, the set $\{A_1, A_2, A_3\}$ is a linearly independent subset of the vector space of (2 x 2) matrices.

17. The given matrices have coordinate vectors $\mathbf{u_1} = [2, 2, 1, 3]^{\mathrm{T}}$,
$\mathbf{u_2} = [1, 4, 0, 5]^{\mathrm{T}}, \mathbf{u_3} = [4, 10, 1, 13]^{\mathrm{T}}$, respectively. The set
$\{\mathbf{u_1}, \mathbf{u_2}, \mathbf{u_3}\}$ is linearly dependent in R^4. For example $-\mathbf{u_1} - 2\mathbf{u_2} + \mathbf{u_3} = \theta$. It follows that $\{A_1, A_2, A_3\}$ is linearly dependent; indeed $-A_1 - 2A_2 + A_3 = \mathcal{O}$.

19. The polynomials $p_1(x), p_2(x), p_3(x)$, have coordinate vectors
$\mathbf{u_1} = [-1, 2, 1, 0]^{\mathrm{T}}, \mathbf{u_2} = [2, -5, 1, 0]^{\mathrm{T}}, \mathbf{u_3} = [0, -1, 3, 0]^{\mathrm{T}}$, respectively. The set $\{\mathbf{u_1} . \mathbf{u_2}, \mathbf{u_3}\}$ is linearly dependent; for example $-2\mathbf{u_1} - \mathbf{u_2} + \mathbf{u_3} = \theta$. It follows that the set $\{p_1(x), p_2(x), p_3(x)\}$ is linearly dependent; indeed $-2p_1(x) - p_2(x) + p_3(x) = 0$.

21. The given polynomials have coordinate vectors $\mathbf{u_1} = [1, 0, 0, 1]^{\mathrm{T}}$,
$\mathbf{u_2} = [1, 0, 1, 0]^{\mathrm{T}}, \mathbf{u_3} = [1, 1, 0, 0]^{\mathrm{T}}, \mathbf{u_4} = [1, 0, 0, 0]^{\mathrm{T}}$, respectively.
Since $\{\mathbf{u_1}, \mathbf{u_2}, \mathbf{u_3}, \mathbf{u_4}\}$ is a linearly independent subset of R^4, the given set of polynomials is a linearly independent subset of $\mathcal{P}_3$.

23. The given polynomials have coordinate vectors $\mathbf{u_1} = [1, 2, 1]^{\mathrm{T}}$,
$\mathbf{u_2} = [2, 5, 0]^{\mathrm{T}}, \mathbf{u_3} = [3, 7, 1]^{\mathrm{T}}, \mathbf{u_4} = [1, 1, 3]^{\mathrm{T}}$, respectively. In the equation $x_1\mathbf{u_1} + x_2\mathbf{u_2} + x_3\mathbf{u_3} + x_4\mathbf{u_4} = \theta$, x_3 and x_4 are arbitrary. It follows that $\{\mathbf{u_1}, \mathbf{u_2}\}$ is a basis for $\mathrm{Sp}\{\mathbf{u_1}, \mathbf{u_2}, \mathbf{u_3}, \mathbf{u_4}\}$. Therefore $\{p_1(x), p_2(x)\}$ is a basis for $\mathrm{Sp}(S)$.

25. The given matrices have coordinate vectors $\mathbf{u_1} = [1, 2, -1, 3]^{\mathrm{T}}$, $\mathbf{u_2} = [-2, 1, 2, -1]^{\mathrm{T}}, \mathbf{u_3} = [-1, -1, 1, -3]^{\mathrm{T}}, \mathbf{u_4} = [-2, 2, 2, 0]^{\mathrm{T}}$, respectively. In the equation $x_1\mathbf{u_1} + x_2\mathbf{u_2} + x_3\mathbf{u_3} + x_4\mathbf{u_4} = \theta$, x_4 is arbitrary so $\{\mathbf{u_1}, \mathbf{u_2}, \mathbf{u_3}\}$ is a basis for $\mathrm{Sp}\{\mathbf{u_1}, \mathbf{u_2}, \mathbf{u_3}, \mathbf{u_4}\}$. It follows that $\{A_1, A_2, A_3\}$ is a basis for $\mathrm{Sp}(S)$.

27. $p(x) = -4p_1(x) + 11p_2(x) - 3p_3(x)$ so $[p(x)]_Q = [-4, 11, -3]^{\mathrm{T}}$.

31. $A = (a + b - 2c + 7d)A_1 + (-b + 2c - 4d)A_2 + (c - 2d)A_3 + dA_4$ so $[A]_Q = [a + b - 2c + 7d, -b + 2c - 4d, c - 2d, d]^{\mathrm{T}}$.

4.5 Dimension

1. (a) We show that V_1 is a subspace. The proof that V_2 is a subspace is similar. Clearly the (3 x 3) zero matrix is lower-triangular so it is in V_1. Now let A and B be in V_1,

$$A = \begin{bmatrix} a_{11} & 0 & 0 \\ a_{21} & a_{22} & 0 \\ a_{31} & a_{32} & a_{33} \end{bmatrix} \text{ and } B = \begin{bmatrix} b_{11} & 0 & 0 \\ b_{21} & b_{22} & 0 \\ b_{31} & b_{32} & b_{33} \end{bmatrix}.$$

$$\text{Then } A + B = \begin{bmatrix} a_{11}+b_{11} & 0 & 0 \\ a_{21}+b_{21} & a_{22}+b_{22} & 0 \\ a_{31}+b_{31} & a_{32}+b_{32} & a_{33}+b_{33} \end{bmatrix},$$

so $A+B$ is in V_1. If c is any scalar then

$$cA = \begin{bmatrix} ca_{11} & 0 & 0 \\ ca_{21} & ca_{22} & 0 \\ ca_{31} & ca_{32} & ca_{33} \end{bmatrix},$$

so cA is in V_1. By Theorem 2 of Section 4.3, V is a subspace of V_1.

(b) The set $\{E_{11}, E_{21}, E_{22}, E_{31}, E_{32}, E_{33}\}$ is a basis for V_1 and $\{E_{11}, E_{12}, E_{13}, E_{22}, E_{23}, E_{33}\}$ is a basis for V_2.

(c) $\dim(V) = 9; \dim(V_1) = 6; \dim(V_2) = 6$.

3. Let $A = [a_{ij}]$ be in $V_1 \cap V_2$. Since A is in $V_1, a_{ij} = 0$ for $i < j$. Likewise A is in V_2 so $a_{ij} = 0$ for $i > j$. Thus $a_{ij} = 0$ if $i \neq j$ and $V_1 \cap V_2$ is the set of (3 x 3) diagonal matrices. $\dim(V_1 \cap V_2) = 3$.

5. The set $\left\{ \begin{bmatrix} 0 & 1 & 0 \\ -1 & 0 & 0 \\ 0 & 0 & 0 \end{bmatrix}, \begin{bmatrix} 0 & 0 & 1 \\ 0 & 0 & 0 \\ -1 & 0 & 0 \end{bmatrix}, \begin{bmatrix} 0 & 0 & 0 \\ 0 & 0 & 1 \\ 0 & -1 & 0 \end{bmatrix} \right\}$ is a basis for W so $\dim(W) = 3$.

7. $p(x) = \sum_{i=0}^{4} a_i x^i$ is in W if and only if $a_0 = 4a_4, a_2 = -5a_4$, a_1, a_3, a_4 arbitrary. A basis for W is $\{x, x^3, 4 - 5x^2 + x^4\}$ so $\dim(W) = 3$.

9. S contains 4 elements and $\dim(\mathcal{P}_2) = 3$.By property (1) of Theorem 8, S is linearly dependent.

11. S contains only two vectors and $\dim(V) = 4$.By property (1) of Theorem 9, S does not span V.

13. The set S contains 5 elements whereas $\dim(V) = 4$. By property (1) of Theorem 8, S is a linearly dependent set.

21. Note that $[\mathbf{w}]_B = [d_1, \ldots, d_n]^{\mathrm{T}}$ and $[\mathbf{w}]_C = [c_1, \ldots, c_n]^{\mathrm{T}}$. Thus
$A[\mathbf{w}]_C = c_1[\mathbf{u_1}]_B + \cdots + c_n[\mathbf{u_n}]_B = [c_1\mathbf{u_1} + \cdots + c_n\mathbf{u_n}]_B = [\mathbf{w}]_B$.

(a) $A = \begin{bmatrix} 1 & -1 & 1 \\ 0 & 1 & -2 \\ 0 & 0 & 1 \end{bmatrix}$.

(b) $[p(x)]_C = [8, 4, 1]^{\mathrm{T}}$ and $[p(x)]_B = A[p(x)]_C = [5, 2, 1]^{\mathrm{T}}$.

23. $A^{-1} = \begin{bmatrix} 1 & 1 & 1 \\ 0 & 1 & 2 \\ 0 & 0 & 1 \end{bmatrix}$.

(a) $p(x) = 6 + 11x + 7x^2$. (b) $p(x) = 4 + 2x - x^2$. (c) $p(x) = 5 + x$.
(d) $p(x) = 8 - 2x - x^2$.

24. $A = \begin{bmatrix} 1 & 0 & 0 & 0 \\ 0 & 1 & 1 & 1 \\ 0 & 0 & 1 & 3 \\ 0 & 0 & 0 & 1 \end{bmatrix}$.

(a) $p(x) = -9 + 4x + x(x-1) + x(x-1)(x-2)$.
(b) $p(x) = -2 + 8x + x(x-1)$.
(c) $p(x) = 1 + x + 3x(x-1) + x(x-1)(x-2)$.
(d) $p(x) = 3 + 5x + 5x(x-1) + x(x-1)(x-2)$.

4.6 Inner-products

1. (1) $<\mathbf{x},\mathbf{x}> = 4x_1^2 + x_2^2 \geq 0$ and $<\mathbf{x},\mathbf{x}> = 0$ if and only if $x_1 = x_2 = 0$.
(2) $<\mathbf{x},\mathbf{y}> = 4x_1y_1 + x_2y_2 = 4y_1x_1 + y_2x_2 = <\mathbf{y},\mathbf{x}>$.
(3) $<a\mathbf{x},\mathbf{y}> = 4ax_1y_1 + ax_2y_2 = a(4x_1y_1 + x_2y_2) = a<\mathbf{x},\mathbf{y}>$.
(4) Let $\mathbf{z} = [z_1, z_2]^{\mathrm{T}}$. Then $<\mathbf{x},\mathbf{y}+\mathbf{z}> = 4x_1(y_1+z_1) + x_2(y_2+z_2) = (4x_1y_1 + x_2y_2) + (4x_1z_1 + x_2z_2) = <\mathbf{x},\mathbf{y}> + <\mathbf{x},\mathbf{z}>$.

3. (1) is immediate since A is positive definite.
(2) $<\mathbf{x},\mathbf{y}> = \mathbf{x}^{\mathrm{T}}A\mathbf{y} = (\mathbf{x}^{\mathrm{T}}A\mathbf{y})^{\mathrm{T}} = \mathbf{y}^{\mathrm{T}}A^{\mathrm{T}}\mathbf{x}^{\mathrm{TT}} =$
$\mathbf{y}^{\mathrm{T}}A\mathbf{x} = <\mathbf{y},\mathbf{x}>$.
(3) $<a\mathbf{x},\mathbf{y}> = (a\mathbf{x})^{\mathrm{T}}A\mathbf{y} = a[\mathbf{x}^{\mathrm{T}}A\mathbf{y}] = a<\mathbf{x},\mathbf{y}>$.

(4) $<\mathbf{x}, \mathbf{y}+\mathbf{z}> = \mathbf{x}^T A(\mathbf{y}+\mathbf{z}) = \mathbf{x}^T \mathbf{A}\mathbf{y} + \mathbf{x}^T \mathbf{A}\mathbf{z} = <\mathbf{x},\mathbf{y}> + <\mathbf{x},\mathbf{z}>$.

5. (1) $<p,p>= a_0^2 + a_1^2 + a_2^2 \geq 0$ with equality if and only if $a_i = 0$ for $0 \leq i \leq 2$.

   (2) $<p,q>= a_0b_0 + a_1b_1 + a_2b_2 = b_0a_0 + b_1a_1 + b_2a_2 = <q,p>$.

   (3) $<ap,q>= aa_0b_0 + aa_1b_1 + aa_2b_2 = a(a_0b_0 + a_1b_1 + a_2b_2) = a<p,q>$.

   (4) Let $r(x) = c_0 + c_1x + c_2x^2$. Then $<p,q+r>= a_0(b_0+c_0) + a_1(b_1+c_1) + a_2(b_2+c_2) = (a_0b_0 + a_1b_1 + a_2b_2) + (a_0c_0 + a_1c_1 + a_2c_2) =$

   $<p,q> + <p,r>$.

7. (1) $<A,A>= a_{11}^2 + a_{12}^2 + a_{21}^2 + a_{22}^2 \geq 0$ with equality if and only if $A = \mathcal{O}$.

   (2) $<A,B>= a_{11}b_{11}+a_{12}b_{12}+a_{21}b_{21}+a_{22}b_{22} = b_{11}a_{11}+b_{12}a_{12}+b_{21}a_{21}+b_{22}a_{22} = <B,A>$.

   (3) $<aA,B>= aa_{11}b_{11}+aa_{12}b_{12}+aa_{21}b_{21}+aa_{22}b_{22} = a(a_{11}b_{11}+a_{12}b_{12}+a_{21}b_{21}+a_{22}b_{22}) = a<A,B>$.

   (4) Let $C = [c_{ij}]$. Then $<A,B+C>= a_{11}(b_{11}+c_{11}) + a_{12}(b_{12}+c_{12}) + a_{21}(b_{21}+c_{21}) + a_{22}(b_{22}+c_{22}) = (a_{11}b_{11}+a_{12}b_{12}+a_{21}b_{21}+a_{22}b_{22}) + (a_{11}c_{11}+a_{12}c_{12}+a_{21}c_{21}+a_{22}c_{22}) = <A,B> + <A,C>$.

9. $<\mathbf{x},\mathbf{y}> = [1,-2]\begin{bmatrix} 1 & 1 \\ 1 & 2 \end{bmatrix}\begin{bmatrix} 0 \\ 1 \end{bmatrix} = -3; \|\mathbf{x}\|^2 = <\mathbf{x},\mathbf{x}> =$

   $\mathbf{x}^T A\mathbf{x} = 5$ so $\|\mathbf{x}\| = \sqrt{5}; \|\mathbf{y}\| = \sqrt{2}; \mathbf{x}-\mathbf{y} = [1,-3]^T$ and

   $\|\mathbf{x}-\mathbf{y}\|^2 = (\mathbf{x}-\mathbf{y})^T A(\mathbf{x}-\mathbf{y}) = 13$. Thus $\|\mathbf{x}-\mathbf{y}\| = \sqrt{13}$.

11. $<p,q>= (-1)1 + (2)2 + 7(7) = 52; \|\mathrm{p}\|^2 = <p,p>= (-1)^2 + 2^2 + 7^2 = 54$ so $\|p\| = 3\sqrt{6}; \|q\|^2 = <q,q>= 1^2 + 2^2 + 7^2 = 54$ so $\|q\| = 3\sqrt{6}; \|p-q\|^2 = <p-q,p-q>= 2^2$ so $\|p-q\| = 2$.

13. For $<\mathbf{x},\mathbf{y}> = \mathbf{x}^T\mathbf{y}$ the graph of S is the circle with equation $x^2 + y^2 = 1$. For $<\mathbf{x},\mathbf{y}> = 4x_1y_1 + x_2y_2$ the graph of S is the ellipse with equation $4x^2 + y^2 = 1$.

15. $\mathbf{v} = a_1\mathbf{u_1} + a_2\mathbf{u_2}$ where $a_1 = <\mathbf{u_1},\mathbf{v}> / <\mathbf{u_1},\mathbf{u_1}> = 7$ and $a_2 = <\mathbf{u_2},\mathbf{v}> / <\mathbf{u_2},\mathbf{u_2}> = 4$.

17. $q(x) = a_0p_0(x) + a_1p_1(x) + a_2p_2(x)$ where $a_0 = <p_0,q> / <p_0,p_0> = -5/3, a_1 = <p_1,q> / <p_1,p_1> = -5$, and

    $a_2 = <p_2,q> / <p_2,p_2> = -4$.

19. $p_0(x) = 1, p_1(x) = x - a_0p_0$ where $a_0 = <x,p_0>/<p_0,p_0> = 0$.

    Thus $p_1(x) = x$. $p_2(x) = x^2 - b_0p_0 - b_1p_1$ where $b_0 =$

$<x^2, p_0>/<p_0, p_0> = 2$ and $b_1 = <x^2, p_1>/<p_1, p_1> = 0$. Thus $p_2(x) = x^2 - 2$. $p_3(x) = x^3 - c_0p_0 - c_1p_1 - c_2p_2$ where $c_0 =$

$<x^3, p_0>/<p_0, p_0> = 0, c_1 = <x^3, p_1>/<p_1, p_1> = 17/5$, and $c_2 = <x^3, p_2>/<p_2, p_2> = 0$. Therefore $p_3(x) = x^3 - (17/5)x$.

$p_4(x) = x^4 - d_0p_0 - d_1p_1 - d_2p_2 - d_3p_3$ where $d_0 =$

$<x^4, p_0> / <p_0, p_0> = 34/5, d_1 = <x^4, p_1> / <p_1, p_1> = 0, d_2 =$

$< x^4, p_2 > / < p_2, p_2 > = 31/7, d_3 = < x^4, p_3 > / < p_3, p_3 > = 0$. Therefore $p_4(x) = x^4 - (31/7)x^2 + 72/35$.

25. From Examples 4 and 5, $p_0(x) = 1, p_1(x) = x - (1/2), p_2(x) = x^2 - x + (1/6), <p_0, p_0> = 1, <p_1, p_1> = 1/12$, and $<p_2, p_2> = 1/180$. Moreover $<x^3, p_0> = 1/4, <x^3, p_1> = 3/40$, and $<x^3, p_2> = 1/120$. By Theorem 13, $p^*(x) = (1/4)p_0(x) +$

   $(9/10)p_1(x) + (3/2)p_2(x) = (3/2)x^2 - (3/5)x + (1/20)$.

27. With $p_3(x)$ as determined in Example 7 and with the calculations done in Example 6 we obtain $p^*(x) \simeq 0.841471p_0(x) -$

   $0.467544p_1(x) - 0.430920p_2(x) + 0.07882p_3(x)$.

29. (a) Clearly $T_0(\cos\theta) = 1 = \cos(0\theta)$ and $T_1(cos\theta) = \cos\theta = \cos(1\theta)$. Suppose we have seen that $T_k(cos\theta) = cos(k\theta)$ for $0 \leq k \leq n$, where $n \geq 1$. Then $T_{n+1}(\cos\theta) = 2\cos\theta\, T_n(\cos\theta) - T_{n-1}(\cos\theta) = 2\cos\theta\cos(n\theta) - \cos(n-1)\theta = \cos(n+1)\theta$ [since $\cos(\alpha+\beta) = 2\cos\alpha\cos\beta - \cos(\alpha-\beta)$].

   (b) $<T_i, T_j> = (2/\pi)\int_{-1}^{1}[T_i(x)T_j(x)/\sqrt{1-x^2}]dx =$
   $-(2/\pi)\int_{\pi}^{0}\cos(i\theta)\cos(j\theta)d\theta = 0$ if $i \neq j$.

   (c) $T_0(x) = 1$ has degree zero and $T_1(x) = x$ has degree one. Suppose $T_k(x)$ has degree k for $0 \leq k \leq n$, where $n \geq 1$. Thus $T_n(x) = a_nx^n + \cdots + a_1x + a_0$ and $T_{n-1}(x) = b_{n-1}x^{n-1} + \cdots + b_1x + b_0$, where $a_n \neq 0$. Using (R) we obtain $T_{n+1}(x) = 2a_nx^{n+1} + \cdots + (2a_0 - b_1)x - b_0$. In particular, $T_{n+1}(x)$ has degree $n+1$. It follows by induction that $T_k(x)$ is a polynomial of degree k for each integer $k \geq 0$.

   (d) $T_2(x) = 2x^2 - 1, T_3(x) = 4x^3 - 3x, T_4(x) = 8x^4 - 8x^2 + 1, T_5(x) = 16x^6 - 20x^3 + 5x$.

4.7 Linear Transformations

1. Let $A = \begin{bmatrix} 1 & 0 \\ 0 & 0 \end{bmatrix}$ and $B = \begin{bmatrix} 0 & 0 \\ 0 & 1 \end{bmatrix}$. Then $T(A+B) =$

$det(A+B) = det(I) = 1$ whereas $T(A) + T(B) = det(A) + det(B) = 0$. Therefore T is not a linear transformation.

3. T is a linear transformation. If A and B are (2 x 2) matrices it is straightforward to see that $tr(A+B) = tr(A) + tr(B)$; thus
$T(A+B) = T(A) + T(B)$. Likewise if c is a scalar, $tr(cA) =$
$c\,tr(A)$ so $T(cA) = cT(A)$.

5. Let f and g be in $C[-1,1]$ and let c be a scalar. Then $T(f+g) = (f+g)'(0) = f'(0) + g'(0) = T(f) + T(g)$, and $T(cf) = (cf)'(0) =$
$cf'(0) = cT(f)$. Therefore T is a linear transformation.

7. T is not a linear transformation. For example $T(1+\theta(x)) = T(1) = 2 + x + x^2$ whereas $T(1) + T(\theta(x)) = (2 + x + x^2) + (1 + x + x^2) = 3 + 2x + 2x^2$.

9. (a) $T(p) = 3T(1) - 2T(x) + 4T(x^2) = 3(1+x^2) - 2(x^2 - x^3) +$
$4(2 + x^3) = 11 + x^2 + 6x^3$.
   (b) $T(a_0 + a_1x + a_2x^2) = a_0T(1) + a_1T(x) + a_2T(x^2) = a_0(1+x^2) + a_1(x^2 - x^3) + a_2(2 + x^3) =$
$(a_0 + 2a_2) + (a_0 + a_1)x^2 + (-a_1 + a_2)x^3$.

11. (a) $T(A) = -2T(E_{11}) + 2T(E_{12}) + 3T(E_{21}) + 4T(E_{22}) = 8 + 14x - 9x^2$.
   (b) $T\left(\begin{bmatrix} a & b \\ c & d \end{bmatrix}\right) = aT(E_{11}) + bT(E_{12}) + cT(E_{21}) + dT(E_{22}) =$
$(a + b + 2d) + (-a + b + 2c + d)x + (b - c - 2d)x^2$.

13. (a) By property 1 of Theorem 15, $\mathcal{R}(T) =$
Sp $\{T(1), T(x), T(x^2), T(x^3), T(x^4)\}$ =Sp$\{0, 0, 2, 6x, 12x^2\}$ =
Sp$\{2, 6x, 12x^2\}$. It follows that rank $(T) = 3$. Since $\mathcal{R}(T) \subseteq \mathcal{P}_2$
and dim $(\mathcal{P}_2) = 3$, we have $\mathcal{R}(T) = \mathcal{P}_2$.
   (b) By property 3 of Theorem 15, nullity $(T) =$
dim $(\mathcal{P}_4) -$ rank $(T) = 5 - 3 = 2$. Since nullity $(T) > 0, T$ is not one to one.
   (c) We wish to determine $q(x) = b_0 + b_1x + b_2x^2 + b_3x^3 + b_4x^4$ in $\mathcal{P}_4$ such that $a_0 + a_1x + a_2x^2 = T(q) = 2b_2 + 6b_3x + 12b_4x^2$.
Equating coefficients gives $b_2 = a_0/2, b_3 = a_1/6, b_4 = a_2/12, b_0$
and b_1 arbitrary. In particular,$q(x) = (a_0/2)x^2 + (a_1/6)x^3 + (a_2/12)x^4$ is one choice.

15. $\mathcal{N}(T) = \{p(x) = a_0 + a_1x + a_2x^2 : a_0 + 2a_1 + 4a_2 = 0\}$. It
follows that nullity $(T) = 2$. Consequently rank $(T) = 1$ and
$\mathcal{R}(T) = R^1$.

17. (a) Let $\mathbf{u}$, $\mathbf{v}$ be vectors in V and let c be a scalar. Then
$I(\mathbf{u} + \mathbf{v}) = \mathbf{u} + \mathbf{v} = I(\mathbf{u}) + I(\mathbf{v})$ and $I(c\mathbf{u}) = c\mathbf{u} = cI(\mathbf{u})$.
Therefore I is a linear transformation.

(b) The vector $\mathbf{v}$ is in $\mathcal{N}(I)$ if and only if $\theta= I(\mathbf{v}) = \mathbf{v}$. Thus $\mathcal{N}(I) = \{\theta\}$. For each $\mathbf{v}$ in $V, I(\mathbf{v}) = \mathbf{v}$ so $\mathcal{R}(I) = V$.

19. Recall that $5 = \dim(\mathcal{P}_4) = \text{rank}(T) + \text{nullity}(T)$. Moreover $\mathcal{R}(T) \subseteq \mathcal{P}_2$ so $\text{rank}(T) \leq 3$. The possibilities are:

$$\begin{array}{rcccc} \text{rank}(T) & 3 & 2 & 1 & 0 \\ \text{nullity}(T) & 2 & 3 & 4 & 5. \end{array}$$

Since $\text{nullity}(T) \geq 2, T$ cannot be one to one.

21. Recall that $3 = \dim(R^3) = \text{rank}(T) + \text{nullity}(T)$. Moreover $\mathcal{N}(T) \subseteq R^3$ so $\text{nullity}(T) \leq 3$. The possibilities are:

$$\begin{array}{rcccc} \text{rank}(T) & 3 & 2 & 1 & 0 \\ \text{nullity}(T) & 0 & 1 & 2 & 3. \end{array}$$

Since $\dim(\mathcal{P}_3) = 4$ and $\text{rank}(T) < 4, \mathcal{R}(T) = \mathcal{P}_3$ is not a possibility.

27. (a) Let A and B be (2 x 2) matrices. Then $T(A + B) = (A + B)^{\mathrm{T}} = A^{\mathrm{T}} + B^{\mathrm{T}} = T(A) + T(B)$. If c is a scalar then $T(cA) = (cA)^{\mathrm{T}} = cA^{\mathrm{T}} = cT(A)$. This proves that T is a linear transformation.

(b) If A is in $\mathcal{N}(T)$ then $T(A) = A^{\mathrm{T}} = \mathcal{O}$. It follows that $A = \mathcal{O}$ and that $\text{nullity}(T) = 0$. Therefore $\text{rank}(T) = 4$. Consequently T is one to one and $\mathcal{R}(T) = V$.

(c) Let B be in V and set $C = B^{\mathrm{T}}$. Then $T(C) = C^{\mathrm{T}} = B^{\mathrm{TT}} = B$. Therefore $\mathcal{R}(T) = V$.

4.8 Operations with Linear Transformations

1. $(S+T)(p) = S(p) + T(p) = p'(0) + (x+2)p(x)$. In particular, $(S+T)(x) = 1 + (x+2)x = x^2 + 2x + 1$ and $(S+T)(x^2) = 0 + (x+2)x^2 = x^3 + 2x^2$.

3. $(H \circ T)(p) = H(T(p)) = H((x+2)p(x)) = [(x+2)p(x)]' + 2p(0) = (x+2)p'(x) + p(x) + 2p(0)$. The domain for $H \circ T$ is $\mathcal{P}_3$ and $(H \circ T)(x) = 2x + 2$.

5. (a) If $p(x) = \sum_{i=0}^{3} a_i x^i$ then $T(p) = 2a_0 + (a_0 + 2a_1)x + (a_1 + 2a_2)x^2 + (a_2 + 2a_3)x^3 + a_3x^4$. In particular $T(p) = \theta(x)$ if and only if $p(x) = \theta(x)$. Therefore T is one to one. Now $\text{rank}(T) = \dim(\mathcal{P}_3) - \text{nullity}(T) = 4$. Since $\mathcal{R}(T) \subseteq \mathcal{P}_4$ and $\dim(\mathcal{P}_4) = 5$, $\mathcal{R}(T) \neq \mathcal{P}_4$; that is, T is not onto.

(b) It is easy to verify that $T(p) = x$ is impossible. Therefore $T^{-1}(x)$ is not defined.

7. Let $p(x) = ae^x + be^{2x} + ce^{3x}$ be in V. Then $T(p(x)) = p'(x) = ae^x + 2be^{2x} + 3ce^{3x}$. Since $B = \{e^x, e^{2x}, e^{3x}\}$ is a linearly independent set, it follows that $T(p(x)) = \theta(x)$ if and only if $a = b = c = 0$. Thus $\mathcal{N}(T) = \{\theta(x)\}$ and T is one to one. The set B is a basis for V so $\dim(V) = 3$. Thus $\operatorname{rank}(T) = \dim(V) - \operatorname{nullity}(T) = 3$. It follows that T is onto. Therefore T is invertible. Moreover $T^{-1}(e^x) = e^x, T^{-1}(e^{2x}) = (1/2)e^{2x}$, and $T^{-1}(e^{3x}) = (1/3)e^{3x}$. This implies that $T^{-1}(ae^x + be^{2x} + ce^{3x}) = ae^x + (b/2)e^{2x} + (c/3)e^{3x}$.

9. If A is in $\mathcal{N}(T)$ then $T(A) = A^T = \mathcal{O}$. Therefore $A = \mathcal{O}$, so $\mathcal{N}(T) = \{\mathcal{O}\}$. It follows that T is one to one. Further $\operatorname{rank}(T) = \dim(V) - \operatorname{nullity}(T) = 4$ so T is onto. Therefore T is invertible. In fact $T^{-1} = T$ since $(A^T)^T = A$.

11. (a) Since $\dim(V) = 4, V$ is isomorphic to R^4 by Theorem 17.

 (b) Since $\dim(\mathcal{P}_3) = 4 = \dim(V)$, V and $\mathcal{P}_3$ are isomorphic by the corollary to Theorem 17.

 (c) It is easily shown that $T : V \to \mathcal{P}_3$ defined by $T\left(\begin{bmatrix} a & b \\ c & d \end{bmatrix}\right) = a + bx + cx^2 + dx^3$ is an isomorphism.

4.9 Matrix Representations for Linear Transformations

1. $S(1) = 0, S(x) = 1, S(x^2) = 0$, and $S(x^3) = 0$. Thus $[S(1)]_C = [S(x^2)]_C = [S(x^3)]_C = [0,0,0,0,0]^{\mathrm{T}}$, while $[S(x)]_C = [1,0,0,0,0]^{\mathrm{T}}$.

The matrix for S is $\begin{bmatrix} 0 & 1 & 0 & 0 \\ 0 & 0 & 0 & 0 \\ 0 & 0 & 0 & 0 \\ 0 & 0 & 0 & 0 \\ 0 & 0 & 0 & 0 \end{bmatrix}$.

3. (a) $(S+T)(1) = 2+x, (S+T)(x) = 1+2x+x^2, (S+T)(x^2) = 2x^2+x^3, (S+T)(x^3) = 2x^3+x^4$. Therefore $[(S+T)(1)]_C = [2,1,0,0,0]^T, [(S+T)(x)]_C = [1,2,1,0,0]^T, [(S+T)](x^2)]_C = [0,0,2,1,0]^T$, and $[(S+T)(x^3)]_C = [0,0,0,2,1]^T$.

The matrix for $S+T$ is the matrix $\begin{bmatrix} 2 & 1 & 0 & 0 \\ 1 & 2 & 0 & 0 \\ 0 & 1 & 2 & 0 \\ 0 & 0 & 1 & 2 \\ 0 & 0 & 0 & 1 \end{bmatrix}$.

(b) By Theorem 19 the matrix for $S+T$ is the sum of matrices for S and T. This is easily verified.

5. $H(1) = 1, H(x) = 1, H(x^2) = 2x, H(x^3) = 3x^2$, and $H(x^4) = 4x^3$. Therefore $[H(1)]_B = [H(x)]_B = [1,0,0,0]^T, [H(x^2)]_B$
$= [0,2,0,0]^T, [H(x^3)]_B = [0,0,3,0]^T$, and $[H(x^4)]_B = [0,0,0,4]^T$.

The matrix for H is the matrix $\begin{bmatrix} 1 & 1 & 0 & 0 & 0 \\ 0 & 0 & 2 & 0 & 0 \\ 0 & 0 & 0 & 3 & 0 \\ 0 & 0 & 0 & 0 & 4 \end{bmatrix}$.

7. (a) $(T \circ H)(1) = 2+x, (T \circ H)(x) = 2+x, (T \circ H)(x^2) = 4x+2x^2, (T \circ H)(x^3) = 6x^2+3x^3, (T \circ H)(x^4) = 8x^3+4x^4$. Therefore
$[(T \circ H)(1)]_C = [(T \circ H)(x)]_C = [2,1,0,0,0]^T, [(T \circ H)(x^2)]_C$
$= [0,4,2,0,0]^T, [(T \circ H)(x^3)]_C = [0,0,6,3,0]^T$, and
$[(T \circ H)(x^4)]_C = [0,0,0,8,4]^T$. Thus the matrix for $T \circ H$ is the matrix

$$\begin{bmatrix} 2 & 2 & 0 & 0 & 0 \\ 1 & 1 & 4 & 0 & 0 \\ 0 & 0 & 2 & 6 & 0 \\ 0 & 0 & 0 & 3 & 8 \\ 0 & 0 & 0 & 0 & 4 \end{bmatrix}.$$

(b) Let D, E, and F denote the matrices for T, H, and $T \circ H$, respectively (cf. Exercises 2, 5, and 7(a)). By Theorem 20, $F = DE$ and it is easily verified that this the case.

9. (a) $[p]_B = [a_0, a_1, a_2, a_3]^T; T(p) = 2a_0 + (a_0+2a_1)x + (a_1+2a_2)x^2 + (a_2+2a_3)x^3 + a_3x^4$ so $[T(p)]_C = [2a_0, a_0+2a_1, a_1+2a_2, a_2+2a_3, a_3]^T$.

11. (a) $Q = \begin{bmatrix} 1 & 0 & 0 \\ 0 & 2 & 0 \\ 0 & 0 & 3 \end{bmatrix}$.

(b) $P = \begin{bmatrix} 1 & 0 & 0 \\ 0 & 1/2 & 0 \\ 0 & 0 & 1/3 \end{bmatrix}$.

(c) Clearly $P = Q^{-1}$.

13. (a) $T(E_{11}) = E_{11}, T(E_{12}) = E_{21}, T(E_{21}) = E_{12}$, and $T(E_{22}) = E_{22}$. Therefore $Q = \begin{bmatrix} 1 & 0 & 0 & 0 \\ 0 & 0 & 1 & 0 \\ 0 & 1 & 0 & 0 \\ 0 & 0 & 0 & 1 \end{bmatrix}$.

(b) If $A = [a_{ij}]$ is a (2 x 2) matrix then $[A]_B = [a_{11}, a_{12}, a_{21}, a_{22}]^{\mathrm{T}}$ whereas $[A^{\mathrm{T}}]_B = [a_{11}, a_{21}, a_{12}, a_{22}]^{\mathrm{T}}$. Clearly $Q[A]_B = [A^{\mathrm{T}}]_B$.

15. $S(x+1) = 3 + 3x - x^2, S(x+2) = 6 + 3x - x^2$, and $S(x^2) = 3x^2$. Therefore $[S(x+1)]_C = [3, 3, -1, 0]^{\mathrm{T}}, [S(x+2)]_C = [6, 3, -1, 0]^{\mathrm{T}}$,
and $[S(x^2)]_C = [0, 0, 3, 0]^{\mathrm{T}}$ and the matrix representation for S
is $\begin{bmatrix} 3 & 6 & 0 \\ 3 & 3 & 0 \\ -1 & -1 & 3 \\ 0 & 0 & 0 \end{bmatrix}$.

17. $T(1) = \begin{bmatrix} 1 \\ 0 \\ 0 \end{bmatrix}, T(x) = \begin{bmatrix} 0 \\ 3 \\ 1 \end{bmatrix}$, and $T(x^2) = \begin{bmatrix} 0 \\ 6 \\ 4 \end{bmatrix}$, so the matrix for T is given by $\begin{bmatrix} 1 & 0 & 0 \\ 0 & 3 & 6 \\ 0 & 1 & 4 \end{bmatrix}$.

19. $T(\mathbf{v_1}) = 0\mathbf{v_1} + 1\mathbf{v_2} + 0\mathbf{v_3} + 0\mathbf{v_4}, T(\mathbf{v_2}) = 0\mathbf{v_1} + 0\mathbf{v_2} + 1\mathbf{v_3} + 0\mathbf{v_4}, T(\mathbf{v_3}) = 1\mathbf{v_1} + 1\mathbf{v_2} + 0\mathbf{v_3} + 0\mathbf{v_4}$,
and $T(\mathbf{v_4}) = 1\mathbf{v_1} + 0\mathbf{v_2} + 0\mathbf{v_3} + 3\mathbf{v_4}$, so the matrix of T is $\begin{bmatrix} 0 & 0 & 1 & 1 \\ 1 & 0 & 1 & 0 \\ 0 & 1 & 0 & 0 \\ 0 & 0 & 0 & 3 \end{bmatrix}$.

21. $T(1) = -4 + 3x - x^2, T(x) = -2 + 3x + 2x^2$, and $T(x^2) = 3x^2$ so the matrix for T is $\begin{bmatrix} -4 & -2 & 0 \\ 3 & 3 & 0 \\ -1 & 2 & 3 \end{bmatrix}$.

23. $T(1-3x+7x^2) = 2(1-3x+7x^2), T(6-3x+2x^2) = -3(6-3x+2x^2)$, and $T(x^2) = 3x^2$.

Therefore the matrix of T is $\begin{bmatrix} 2 & 0 & 0 \\ 0 & -3 & 0 \\ 0 & 0 & 3 \end{bmatrix}$.

31. If Q is the matrix for T then $T(1) = 1-x^2, T(x) = x+x^2, T(x^2) = 2$, and $T(x^3) = x-x^2$. Therefore $T(a_0+a_1x+a_2x^2+a_3x^3) = a_0T(1)+a_1(T(x)+a_2T(x^2)+a_3T(x^3) = (a_0+2a_2)+(a_1+a_3)x+(-a_0+a_1-a_3)x^2$.

4.10 Change of Basis and Diagonalization

1. $T(\mathbf{u_1}) = \mathbf{u_1}$ and $T(\mathbf{u_2}) = 3\mathbf{u_2}$. Therefore $\mathbf{u_1}$ and $\mathbf{u_2}$ are eigenvectors for T corresponding to the eigenvalues $\lambda_1 = 1$ and $\lambda_2 = 3$, respectively. The matrix of T with respect to C is $\begin{bmatrix} 1 & 0 \\ 0 & 3 \end{bmatrix}$.

3. $T(A_1) = 2A_1, T(A_2) = -2A_2, T(A_3) = 3A_3$, and $T(A_4) = -3A_4$.

Therefore A_1, A_2, A_3 and A_4 are eigenvectors for T corresponding to the eigenvalues $\lambda_1 = 2, \lambda_2 = -2, \lambda_3 = 3$, and $\lambda_4 = -3$, respectively. The matrix for T with respect to C is given by $\begin{bmatrix} 2 & 0 & 0 & 0 \\ 0 & -2 & 0 & 0 \\ 0 & 0 & 3 & 0 \\ 0 & 0 & 0 & -3 \end{bmatrix}$.

5. The transition matrix is the matrix $P = \begin{bmatrix} 1 & -1 & -1 \\ 1 & -1 & 0 \\ -1 & 2 & 1 \end{bmatrix}$.

Now $[p(x)]_B = [2,1,0]^{\mathrm{T}}$ and $P[p(x)]_B = [1,1,0]^{\mathrm{T}} = [p(x)]_C$. Denote the polynomials in C by $g_1(x), g_2(x), g_3(x)$, respectively. It follows that $p(x) = g_1(x) + g_2(x)$. Similarly $s(x) = -2g_1(x) - g_2(x) + 2g_3(x), q(x) = -5g_1(x) - 3g_2(x) + 7g_3(x)$, and $r(x) =$ $(a_0 - a_1 - a_2)g_1(x) + (a_0 - a_1)g_2(x) + (-a_0 + 2a_1 + a_2)g_3(x)$.

7. Since $\mathbf{u_1} = (1/3)\mathbf{w_1} + (1/3)\mathbf{w_2}$ and $\mathbf{u_2} = (5/3)\mathbf{w_1} - (1/3)\mathbf{w_2}$, the transition matrix is $P = \begin{bmatrix} 1/3 & 5/3 \\ 1/3 & -1/3 \end{bmatrix}$.

9. The transition matrix is $P = \begin{bmatrix} -1 & 1 & 2 & 3 \\ 1 & 0 & 0 & -3 \\ 0 & 0 & 1 & 0 \\ 0 & 0 & 0 & 1 \end{bmatrix}$.

Since $[p(x)]_B = [2,-7,1,0]^{\mathrm{T}}, P[p(x)]_B = [-7,2,1,0]^{\mathrm{T}} = [p(x)]_C$. Let the polynomials in

C be denoted by $g_1(x), g_2(x), g_3(x)$, and $g_4(x)$, respectively. It follows that $p(x) = -7g_1(x) + 2g_2(x) + g_3(x)$. Similarly $q(x) = 13g_1(x) - 4g_2(x) + g_4(x)$ and $r(x) = -7g_1(x) + 3g_2(x) - 2g_3(x) + g_4(x)$.

11. Note that $T(\mathbf{e_1}) = [2,1]^T = 2\mathbf{e_1} + \mathbf{e_2}$ and $T(\mathbf{e_2}) = [1,2]^T = \mathbf{e_1} + 2\mathbf{e_2}$. Therefore the matrix of T with respect to B is the matrix $Q_1 = \begin{bmatrix} 2 & 1 \\ 1 & 2 \end{bmatrix}$. The transition matrix from B to C is the matrix $P = \begin{bmatrix} -1 & 1 \\ 1 & 1 \end{bmatrix}$ and $P^{-1} = \begin{bmatrix} -1/2 & 1/2 \\ 1/2 & 1/2 \end{bmatrix}$. By Theorem 24 the matrix of T with respect to C is the matrix Q_2 given by $Q_2 = P^{-1}Q_1P = \begin{bmatrix} 1 & 0 \\ 0 & 3 \end{bmatrix}$.

13. The matrix of T with respect to B is $Q_1 = \begin{bmatrix} -3 & 0 & 0 & 5 \\ 0 & 3 & -5 & 0 \\ 0 & 0 & -2 & 0 \\ 0 & 0 & 0 & 2 \end{bmatrix}$. The transition matrix from C to B is $P = \begin{bmatrix} 1 & 0 & 0 & 1 \\ 0 & 1 & 1 & 0 \\ 0 & 1 & 0 & 0 \\ 1 & 0 & 0 & 0 \end{bmatrix}$ and $P^{-1} = \begin{bmatrix} 0 & 0 & 0 & 1 \\ 0 & 0 & 1 & 0 \\ 0 & 1 & -1 & 0 \\ 1 & 0 & 0 & -1 \end{bmatrix}$. By Theorem 24 the matrix of T with respect to C is the matrix $Q_2 = P^{-1}Q_1P = \begin{bmatrix} 2 & 0 & 0 & 0 \\ 0 & -2 & 0 & 0 \\ 0 & 0 & 3 & 0 \\ 0 & 0 & 0 & -3 \end{bmatrix}$.

15. (a) $T(1) = 1, T(x) = 1 + 2x$, and $T(x^2) = 4x + 3x^2$.

Therefore $Q = \begin{bmatrix} 1 & 1 & 0 \\ 0 & 2 & 4 \\ 0 & 0 & 3 \end{bmatrix}$.

(b) Q has characteristic polynomial $p(t) = -(t-1)(t-2)(t-3)$.

Therefore Q has eigenvalues $\lambda_1 = 1, \lambda_2 = 2, \lambda_3 = 3$. The corresponding eigenvectors are $\mathbf{u_1} = [1,0,0]^T, \mathbf{u_2} = [1,1,0]^T$ and $\mathbf{u_3} = [2,4,1]^T$, respectively. If $S = [\mathbf{u_1}, \mathbf{u_2}, \mathbf{u_3}]$ then $S^{-1}QS = R$

where $R = \begin{bmatrix} 1 & 0 & 0 \\ 0 & 2 & 0 \\ 0 & 0 & 3 \end{bmatrix}$.

(c) $C = \{\mathbf{v_1}, \mathbf{v_2}, \mathbf{v_3}\}$ where $[\mathbf{v_i}]_B = \mathbf{u_i}$, the ith column of S. Thus $[\mathbf{v_1}]_B = [1,0,0]^T$ so $\mathbf{v_1} = 1$; $[\mathbf{v_2}]_B = [1,1,0]^T$ so $\mathbf{v_2} =$
$1 + x$; $[\mathbf{v_3}]_B = [2,4,1]^T$ so $\mathbf{v_3} = 2 + 4x + x^2$.

(d) $1 = \mathbf{v_1}$, $x = -\mathbf{v_1} + \mathbf{v_2}$, and $x^2 = 2\mathbf{v_1} - 4\mathbf{v_2} + \mathbf{v_3}$ so the transition matrix is

$$P = \begin{bmatrix} 1 & -1 & 2 \\ 0 & 1 & -4 \\ 0 & 0 & 1 \end{bmatrix}.$$

(e) $[\mathbf{w_1}]_B = [-8, 7, 1]^T$ and $[\mathbf{w_1}]_C = P[\mathbf{w_1}]_B = [-13, 3, 1]^T$. Therefore $[T(\mathbf{w_1})]_C = R[\mathbf{w_1}]_C = [-13, 6, 3]^T$. It follows that $T(\mathbf{w_1}) = -13\mathbf{v_1} + 6\mathbf{v_2} + 3\mathbf{v_3} = -1 + 18x + 3x^2$. Similarly $[T(\mathbf{w_2})]_C = [7, -8, 3]^T$ so $T(\mathbf{w_2}) = 7\mathbf{v_1} - 8\mathbf{v_2} + 3\mathbf{v_3} = 5 + 4x + 3x^2$.
Finally, $[T(\mathbf{w_3})]_C = [11, -22, 6]^T$ so $T(\mathbf{w_3}) = 11\mathbf{v_1} - 22\mathbf{v_2} + 6\mathbf{v_3} = 1 + 2x + 6x^2$.

Chapter 5

Determinants

5.1 Introduction (No exercises)

5.2 Cofactor Expansion of Determinants

1. $\det(A) = 1(1) - 3(2) = -5$.

3. $\det(A) = 2(8) - 4(4) = 0; \mathbf{x} = a\begin{bmatrix} -2 \\ 1 \end{bmatrix}, a \neq 0$.

5. $\det(A) = 4(7) - 3(1) = 25$.

7. $\det(A) = 4(1) - 1(-2) = 6$.

9. $A_{11} = (-1)^2\begin{vmatrix} 1 & 3 \\ 1 & 1 \end{vmatrix} = -2; A_{12} = (-1)^3\begin{vmatrix} 0 & 3 \\ 2 & 1 \end{vmatrix} = 6;$

$A_{13} = (-1)^4\begin{vmatrix} 0 & 1 \\ 2 & 1 \end{vmatrix} = -2; A_{33} = (-1)^6\begin{vmatrix} 1 & 2 \\ 0 & 1 \end{vmatrix} = 1.$

11. $A_{11} = \begin{vmatrix} 2 & 2 \\ 2 & 1 \end{vmatrix} = -2; A_{12} = -\begin{vmatrix} -1 & 2 \\ 3 & 1 \end{vmatrix} = 7;$

$A_{13} = \begin{vmatrix} -1 & 2 \\ 3 & 2 \end{vmatrix} = -8; A_{33} = \begin{vmatrix} 2 & -1 \\ -1 & 2 \end{vmatrix} = 3.$

13. $A_{11} = \begin{vmatrix} 1 & 0 \\ 1 & 3 \end{vmatrix} = 3; A_{12} = -\begin{vmatrix} 2 & 0 \\ 0 & 3 \end{vmatrix} = -6;$

$A_{13} = \begin{vmatrix} 2 & 1 \\ 0 & 1 \end{vmatrix} = 2; A_{33} = \begin{vmatrix} -1 & 1 \\ 2 & 1 \end{vmatrix} = -3.$

15. $\det(A) = A_{11} + 2A_{12} + A_{13} = -2 + 2(6) + (-2) = 8.$

17. $\det(A) = 2A_{11} - A_{12} + 3A_{13} = 2(-2) - 7 + 3(-8) = -35.$

19. $\det(A) = -A_{11} + A_{12} - A_{13} = -3 - 6 - 2 = -11.$

21. $\det(A) = 2\begin{vmatrix} 0 & 0 & 1 \\ 1 & 2 & 0 \\ 1 & 1 & 2 \end{vmatrix} + (-1)\begin{vmatrix} 3 & 0 & 1 \\ 2 & 2 & 0 \\ 3 & 1 & 2 \end{vmatrix} + (-1)\begin{vmatrix} 3 & 0 & 1 \\ 2 & 1 & 0 \\ 3 & 1 & 2 \end{vmatrix} +$

$2(-1)\begin{vmatrix} 3 & 0 & 0 \\ 2 & 1 & 2 \\ 3 & 1 & 1 \end{vmatrix} = 2\begin{vmatrix} 1 & 2 \\ 1 & 1 \end{vmatrix} + (-1)\left[3\begin{vmatrix} 2 & 0 \\ 1 & 2 \end{vmatrix} + \begin{vmatrix} 2 & 2 \\ 3 & 1 \end{vmatrix}\right] +$

$(-1)\left[3\begin{vmatrix} 1 & 0 \\ 1 & 2 \end{vmatrix} + \begin{vmatrix} 2 & 1 \\ 3 & 1 \end{vmatrix}\right] + (-2)(3)\begin{vmatrix} 1 & 2 \\ 1 & 1 \end{vmatrix} = -9.$

23. $\det(A) = 2\begin{vmatrix} 3 & 1 & 2 \\ 1 & 2 & 1 \\ 3 & 1 & 4 \end{vmatrix} + 2\begin{vmatrix} 1 & 3 & 2 \\ 0 & 1 & 1 \\ 0 & 3 & 4 \end{vmatrix} =$

$2\left[3\begin{vmatrix} 2 & 1 \\ 1 & 4 \end{vmatrix} - \begin{vmatrix} 1 & 1 \\ 3 & 4 \end{vmatrix} + 2\begin{vmatrix} 1 & 2 \\ 3 & 1 \end{vmatrix}\right] +$

$2\left[\begin{vmatrix} 1 & 1 \\ 3 & 4 \end{vmatrix} - 3\begin{vmatrix} 0 & 1 \\ 0 & 4 \end{vmatrix} + 2\begin{vmatrix} 0 & 1 \\ 0 & 3 \end{vmatrix}\right] = 22.$

25. $\det(A) = a_{11}A_{11} + a_{12}A_{12} + a_{13}A_{13} = (1)(10) + (3)(5) + (2)(-10) = 5$; $a_{21}A_{21} + a_{22}A_{22} + a_{23}A_{23} = (-1)(-5) + (4)(-1) + (1)(4) = 5$; $a_{31}A_{31} + a_{32}A_{32} + a_{33}A_{33} = (2)(-5) + (2)(-3) + (3)(7) = 5$.

27. $a_{11}A_{21} + a_{12}A_{22} + a_{13}A_{23} = (1)(-5) + (3)(-1) + (2)(4) = 0$; $a_{11}A_{31} + a_{12}A_{32} + a_{13}A_{33} = (1)(-5) + (3)(-1) + (2)(7) = 0$.

29. $C = \begin{bmatrix} 10 & 5 & -10 \\ -5 & -1 & 4 \\ -5 & -3 & 7 \end{bmatrix}$, $C^T A = \begin{bmatrix} 5 & 0 & 0 \\ 0 & 5 & 0 \\ 0 & 0 & 5 \end{bmatrix} = [\det(A)]I$.
So $A^{-1} = (1/5)C^T = [1/\det(A)]C^T$.

31. $\det(A) = -a_{12}\begin{vmatrix} 0 & a_{23} \\ 0 & a_{33} \end{vmatrix} + a_{13}\begin{vmatrix} 0 & a_{22} \\ 0 & a_{32} \end{vmatrix} = 0.$

33. $A^{\mathrm{T}} = \begin{bmatrix} a_{11} & a_{21} \\ a_{12} & a_{22} \end{bmatrix}$ so $\det(A^{\mathrm{T}}) = a_{11}a_{22} - a_{21}a_{12} = \det(A)$.

35. (a) For $n = 3$ and $n = 4$, $H(n) = n!/2$. For some integer $k \geq 4$ suppose we have seen that $H(k) = k!/2$. If A is a $((k+1) \times (k+1))$ matrix then $\det(A)$ can be obtained by evaluating $k+1$ $(k \times k)$ determinants. Thus the number of (2 x 2) determinants in the expansion of $\det(A)$ is $(k+1)H(k) = (k+1)!/2$. It follows by induction that $H(n) = n!/2$ for every positive integer $n, n \geq 2$.

(b) Note that evaluating a single (2 x 2) determinant requires 3 operations, two multiplications and one subtraction.

n	$H(n)$	Time required
2	1	3 seconds
5	60	3 minutes
10	1,814,400	1512 hours.

5.3 Elementary Operations and Determinants

1. $$\begin{vmatrix} 1 & 2 & 1 \\ 2 & 0 & 1 \\ 1 & -1 & 1 \end{vmatrix} \overset{\substack{C_2 - 2C_1 \\ C_3 - C_1}}{=} \begin{vmatrix} 1 & 0 & 0 \\ 2 & -4 & -1 \\ 1 & -3 & 0 \end{vmatrix} = \begin{vmatrix} -4 & -1 \\ -3 & 0 \end{vmatrix} = -3.$$

3. $$\begin{vmatrix} 0 & 1 & 2 \\ 3 & 1 & 2 \\ 2 & 0 & 3 \end{vmatrix} \overset{C_1 \leftrightarrow C_2}{=} - \begin{vmatrix} 1 & 0 & 2 \\ 1 & 3 & 2 \\ 0 & 2 & 3 \end{vmatrix} \overset{C_3 - 2C_1}{=} - \begin{vmatrix} 1 & 0 & 0 \\ 1 & 3 & 0 \\ 0 & 2 & 3 \end{vmatrix} =$$
$$- \begin{vmatrix} 3 & 0 \\ 2 & 3 \end{vmatrix} = -9.$$

5. $$\begin{vmatrix} 0 & 1 & 3 \\ 2 & 1 & 2 \\ 1 & 1 & 2 \end{vmatrix} \overset{C_1 \leftrightarrow C_2}{=} - \begin{vmatrix} 1 & 0 & 3 \\ 1 & 2 & 2 \\ 1 & 1 & 2 \end{vmatrix} \overset{C_3 - 3C_1}{=}$$
$$- \begin{vmatrix} 1 & 0 & 0 \\ 1 & 2 & -1 \\ 1 & 1 & -1 \end{vmatrix} = - \begin{vmatrix} 2 & -1 \\ 1 & -1 \end{vmatrix} = 1.$$

7. $\det(B) = -2\det(A) = -6.$

9. $\det(B) = \det(A) = 3.$

11. $\det(B) = \det(A) = 3.$

13. $$\begin{vmatrix} 1 & 0 & 0 & 0 \\ 2 & 0 & 0 & 3 \\ 1 & 1 & 0 & 1 \\ 1 & 4 & 2 & 2 \end{vmatrix} \overset{C_2 \leftrightarrow C_4}{=} - \begin{vmatrix} 1 & 0 & 0 & 0 \\ 2 & 3 & 0 & 0 \\ 1 & 1 & 0 & 1 \\ 1 & 2 & 2 & 4 \end{vmatrix} \overset{C_3 \leftrightarrow C_4}{=}$$

$$\begin{vmatrix} 1 & 0 & 0 & 0 \\ 2 & 3 & 0 & 0 \\ 1 & 1 & 1 & 0 \\ 1 & 2 & 4 & 2 \end{vmatrix} = 6.$$

15. $$\begin{vmatrix} 0 & 1 & 0 & 0 \\ 0 & 2 & 0 & 3 \\ 2 & 1 & 0 & 6 \\ 3 & 2 & 2 & 4 \end{vmatrix} \begin{matrix} C_1 \leftrightarrow C_2 \\ C_3 \leftrightarrow C_4 \\ = \end{matrix} \begin{vmatrix} 1 & 0 & 0 & 0 \\ 2 & 0 & 3 & 0 \\ 1 & 2 & 6 & 0 \\ 2 & 3 & 4 & 2 \end{vmatrix} \begin{matrix} C_2 \leftrightarrow C_3 \\ = \end{matrix}$$

$$-\begin{vmatrix} 1 & 0 & 0 & 0 \\ 2 & 3 & 0 & 0 \\ 1 & 6 & 2 & 0 \\ 2 & 4 & 3 & 2 \end{vmatrix} = -12.$$

17. $$\begin{vmatrix} 2 & 4 & -2 & -2 \\ 1 & 3 & 1 & 2 \\ 1 & 3 & 1 & 3 \\ -1 & 2 & 1 & 2 \end{vmatrix} \begin{matrix} C_2 - 2C_1 \\ C_3 + C_2 \\ C_4 + C_1 \\ = \end{matrix} \begin{vmatrix} 2 & 0 & 0 & 0 \\ 1 & 1 & 2 & 3 \\ 1 & 1 & 2 & 4 \\ -1 & 4 & 0 & 1 \end{vmatrix} \begin{matrix} C_3 - 2C_2 \\ C_4 - 3C_2 \\ = \end{matrix}$$

$$\begin{vmatrix} 2 & 0 & 0 & 0 \\ 1 & 1 & 0 & 0 \\ 1 & 1 & 0 & 1 \\ -1 & 4 & -8 & -11 \end{vmatrix} = 2\begin{vmatrix} 0 & 1 \\ -8 & -11 \end{vmatrix} = 16.$$

19. $$\begin{vmatrix} 1 & 2 & 0 & 3 \\ 2 & 5 & 1 & 1 \\ 2 & 0 & 4 & 3 \\ 0 & 1 & 6 & 2 \end{vmatrix} \begin{matrix} R_2 - 2R_1 \\ R_3 - 2R_1 \\ = \end{matrix} \begin{vmatrix} 1 & 2 & 0 & 3 \\ 0 & 1 & 1 & -5 \\ 0 & -4 & 4 & -3 \\ 0 & 1 & 6 & 2 \end{vmatrix} \begin{matrix} R_3 + 4R_4 \\ R_4 - R_2 \\ = \end{matrix}$$

$$\begin{vmatrix} 1 & 2 & 0 & 3 \\ 0 & 1 & 1 & -5 \\ 0 & 0 & 8 & -23 \\ 0 & 0 & 5 & 7 \end{vmatrix} = \begin{vmatrix} 8 & -23 \\ 5 & 7 \end{vmatrix} = 171.$$

21. $$\begin{vmatrix} 1 & 1 & 2 & 1 \\ 0 & 1 & 4 & 1 \\ 2 & 1 & 3 & 0 \\ 2 & 2 & 1 & 2 \end{vmatrix} \begin{matrix} R_3 - 2R_1 \\ R_4 - 2R_1 \\ = \end{matrix} \begin{vmatrix} 1 & 1 & 2 & 1 \\ 0 & 1 & 4 & 1 \\ 0 & -1 & -1 & -2 \\ 0 & 0 & -3 & 0 \end{vmatrix} \begin{matrix} R_3 + R_2 \\ = \end{matrix}$$

$$\begin{vmatrix} 1 & 1 & 2 & 1 \\ 0 & 1 & 4 & 1 \\ 0 & 0 & 3 & -1 \\ 0 & 0 & -3 & 0 \end{vmatrix} = \begin{vmatrix} 3 & -1 \\ -3 & 0 \end{vmatrix} = -3.$$

23. $\begin{vmatrix} a+1 & a+4 & a+7 \\ a+2 & a+5 & a+8 \\ a+3 & a+6 & a+9 \end{vmatrix} \left\{\begin{matrix} C_3 - C_2 \\ = \\ C_2 - C_1 \end{matrix}\right\} \begin{vmatrix} a+1 & 3 & 3 \\ a+2 & 3 & 3 \\ a+3 & 3 & 3 \end{vmatrix} = 0;$

$\begin{vmatrix} a & 4a & 7a \\ 2a & 5a & 8a \\ 3a & 6a & 9a \end{vmatrix} \left\{\begin{matrix} C_3 - C_2 \\ = \\ C_2 - C_1 \end{matrix}\right\} \begin{vmatrix} a & 3a & 3a \\ 2a & 3a & 3a \\ 3a & 3a & 3a \end{vmatrix} = 0;$

$\begin{vmatrix} a & a^4 & a^7 \\ a^2 & a^5 & a^8 \\ a^3 & a^6 & a^9 \end{vmatrix} = (a)(a^4)(a^7) \begin{vmatrix} 1 & 1 & 1 \\ a & a & a \\ a^2 & a^2 & a^2 \end{vmatrix} = 0.$

31. $\begin{vmatrix} 1 & a & a^2 & a^3 \\ 1 & b & b^2 & b^3 \\ 1 & c & c^2 & c^3 \\ 1 & d & d^2 & d^3 \end{vmatrix} \begin{matrix} C_4 - aC_3 \\ C_3 - aC_2 \\ C_2 - aC_1 \\ = \end{matrix} \begin{vmatrix} 1 & 0 & 0 & 0 \\ 1 & b-a & b(b-a) & b^2(b-a) \\ 1 & c-a & c(c-a) & c^2(c-a) \\ 1 & d-a & d(d-a) & d^2(d-a) \end{vmatrix} =$

$(b-a)(c-a)(d-a) \begin{vmatrix} 1 & b & b^2 \\ 1 & c & c^2 \\ 1 & d & d^2 \end{vmatrix} =$

$(b-a)(c-a)(d-a)(c-b)(d-b)(d-c).$

32. Write $A = [\mathbf{A_1}, \mathbf{A_2}, \ldots, \mathbf{A_n}]$. Then $cA = [c\mathbf{A_1}, c\mathbf{A_2}, \ldots, c\mathbf{A_n}]$. By Theorem 3, $\det(cA) = c \det[\mathbf{A_1}, c\mathbf{A_2}, \ldots, c\mathbf{A_n}] =$ $c^2 \det[\mathbf{A_1}, \mathbf{A_2}, \ldots, c\mathbf{A_n}] = \cdots = c^n \det[\mathbf{A_1}, \mathbf{A_2}, \ldots, \mathbf{A_n}] =$ $c^n \det(A)$.

5.4 Cramer's Rule

1. $\begin{vmatrix} 0 & 1 & 3 \\ 1 & 2 & 1 \\ 3 & 4 & 1 \end{vmatrix} \overset{\{C_1 \leftrightarrow C_2\}}{=} -\begin{vmatrix} 1 & 0 & 3 \\ 2 & 1 & 1 \\ 4 & 3 & 1 \end{vmatrix} \overset{\{C_3 - 3C_1\}}{=}$

$-\begin{vmatrix} 1 & 0 & 0 \\ 2 & 1 & -5 \\ 4 & 3 & -11 \end{vmatrix} \overset{\{C_3 + 5C_2\}}{=} -\begin{vmatrix} 1 & 0 & 0 \\ 2 & 1 & 0 \\ 4 & 3 & 4 \end{vmatrix} = -4.$

3. $\begin{vmatrix} 2 & 2 & 4 \\ 1 & 3 & 4 \\ -1 & 2 & 1 \end{vmatrix} \left\{\begin{matrix} C_2 - C_1 \\ C_3 - 2C_1 \end{matrix}\right\} \underset{}{=} \begin{vmatrix} 2 & 0 & 0 \\ 1 & 2 & 2 \\ -1 & 3 & 3 \end{vmatrix} \overset{\{C_3 - C_2\}}{=}$

$\begin{vmatrix} 2 & 0 & 0 \\ 1 & 2 & 0 \\ -1 & 3 & 0 \end{vmatrix} = 0.$

5. $\begin{vmatrix} 1 & 0 & -2 \\ 3 & 1 & 3 \\ 0 & 1 & 2 \end{vmatrix} \overset{\{C_3+2C_1\}}{=} \begin{vmatrix} 1 & 0 & 0 \\ 3 & 1 & 9 \\ 0 & 1 & 2 \end{vmatrix} \overset{\{C_3-9C_2\}}{=} \begin{vmatrix} 1 & 0 & 0 \\ 3 & 1 & 0 \\ 0 & 1 & -7 \end{vmatrix}$

$\overset{\{(-1/7)C_3\}}{=} -7\begin{vmatrix} 1 & 0 & 0 \\ 3 & 1 & 0 \\ 0 & 1 & 1 \end{vmatrix} \overset{\{C_2-C_3\}}{=}$

$-7\begin{vmatrix} 1 & 0 & 0 \\ 3 & 1 & 0 \\ 0 & 0 & 1 \end{vmatrix} \overset{\{C_1-3C_2\}}{=} -7\begin{vmatrix} 1 & 0 & 0 \\ 0 & 1 & 0 \\ 0 & 0 & 1 \end{vmatrix} = -7.$

7. (a) $\det(AB) = \det(A)\det(B) = (2)(3) = 6$ (b) $\det(AB^2) = \det(A)[\det(B)]^2 = (2)(9) = 18$ (c) $\det(A^{-1}B) = \det(B)/\det(A) = 3/2$ (d) $\det(2A^{-1}) = 8\det(A^{-1}) = 8/\det(A) = 4$ (e) $\det(2A)^{-1} = \det((1/2)A^{-1}) = (1/8)\det(A^{-1}) = 1/[8\det(A)] = 1/16$.

9. $\det(B(\lambda)) = 2\lambda - \lambda^2 = \lambda(2-\lambda)$; $B(\lambda)$ is singular provided $\lambda = 0$ or $\lambda = 2$.

11. $\det(B(\lambda)) = 4 - \lambda^2$; $B(\lambda)$ is singular for $\lambda = \pm 2$.

13. $\det(B(\lambda)) = (\lambda-1)^2(\lambda+2)$; $B(\lambda)$ is singular provided $\lambda = 1$ or $\lambda = -2$.

15. $\det(A) = \begin{vmatrix} 1 & 1 \\ 1 & -1 \end{vmatrix} = -2$; $\det(B_1) = \begin{vmatrix} 3 & 1 \\ -1 & -1 \end{vmatrix} = -2$;

$\det(B_2) = \begin{vmatrix} 1 & 3 \\ 1 & -1 \end{vmatrix} = -4.$

$x_1 = \det(B_1)/\det(A) = 1$; $x_2 = \det(B_2)/\det(A) = 2$.

17. $\det(A) = \begin{vmatrix} 1 & -2 & 1 \\ 1 & 0 & 1 \\ 1 & -2 & 0 \end{vmatrix} = -2$; $\det(B_1) = \begin{vmatrix} -1 & -2 & 1 \\ 3 & 0 & 1 \\ 0 & -2 & 0 \end{vmatrix} = -8$;

$\det(B_2) = \begin{vmatrix} 1 & -1 & 1 \\ 1 & 3 & 1 \\ 1 & 0 & 0 \end{vmatrix} = -4$; $\det(B_3) = \begin{vmatrix} 1 & -2 & -1 \\ 1 & 0 & 3 \\ 1 & -2 & 0 \end{vmatrix} = 2.$

$x_1 = \det(B_1)/\det(A) = 4$; $x_2 = \det(B_2)/\det(A) = 2$;

$x_3 = \det(B_3)/\det(A) = -1$.

19. $\det(A) = \begin{vmatrix} 1 & 1 & 1 & -1 \\ 0 & 1 & -1 & 1 \\ 0 & 0 & 1 & -1 \\ 0 & 0 & 1 & 2 \end{vmatrix} = 3$; $\det(B_1) = \begin{vmatrix} 2 & 1 & 1 & -1 \\ 1 & 1 & -1 & 1 \\ 0 & 0 & 1 & -1 \\ 3 & 0 & 1 & 2 \end{vmatrix} = 3$;

$$\det(B_2)\begin{vmatrix} 1 & 2 & 1 & -1 \\ 0 & 1 & -1 & 1 \\ 0 & 0 & 1 & -1 \\ 0 & 3 & 1 & 2 \end{vmatrix} = 3; \det(B_3) = \begin{vmatrix} 1 & 1 & 2 & -1 \\ 0 & 1 & 1 & 1 \\ 0 & 0 & 0 & -1 \\ 0 & 0 & 3 & 2 \end{vmatrix} = 3;$$

$$\det(B_4) = \begin{vmatrix} 1 & 1 & 1 & 2 \\ 0 & 1 & -1 & 1 \\ 0 & 0 & 1 & 0 \\ 0 & 0 & 1 & 3 \end{vmatrix} = 3.$$

$x_1 = \det(B_1)/\det(A) = 1; x_2 = \det(B_2)/\det(A) = 1;$

$x_3 = \det(B_3)/\det(A) = 1; x_4 = \det(B_4)/\det(A) = 1.$

21. $\det(A) = \begin{vmatrix} 1 & 1 & 1 \\ 0 & 1 & 1 \\ 0 & 0 & 1 \end{vmatrix} = 1; \det(B_1) = \begin{vmatrix} a & 1 & 1 \\ b & 1 & 1 \\ c & 0 & 1 \end{vmatrix} = a - b;$

$$\det(B_2) = \begin{vmatrix} 1 & a & 1 \\ 0 & b & 1 \\ 0 & c & 1 \end{vmatrix} = b - c; \det(B_3) = \begin{vmatrix} 1 & 1 & a \\ 0 & 1 & b \\ 0 & 0 & c \end{vmatrix} = c.$$

$x_1 = a - b, x_2 = b - c, x_3 = c.$

25. $\det(B) = \det(SAS^{-1}) = \det(S)\det(A)\det(S^{-1}) =$
$\det(S)\det(S)^{-1}\det(A) = \det(A).$

27. $\det(A^5) = \det(A)^5 = 3^5 = 243.$

29. $\det(Q) = -33 = (-3)(11) = \begin{vmatrix} 1 & 2 \\ 2 & 1 \end{vmatrix} \begin{vmatrix} 1 & 2 & 2 \\ 3 & 5 & 1 \\ 1 & 4 & 1 \end{vmatrix}.$

5.5 Applications of Determinants

1. $\begin{vmatrix} 1 & 2 & 1 \\ 2 & 3 & 2 \\ -1 & 4 & 1 \end{vmatrix} \overset{\left\{ \begin{array}{c} R_2 - 2R_1 \\ R_3 + R_1 \end{array} \right\}}{=} \begin{vmatrix} 1 & 2 & 1 \\ 0 & -1 & 0 \\ 0 & 6 & 2 \end{vmatrix} \overset{\{R_3 + 6R_2\}}{=}$

$$\begin{vmatrix} 1 & 2 & 1 \\ 0 & -1 & 0 \\ 0 & 0 & 2 \end{vmatrix} = -2.$$

3. $\begin{vmatrix} 0 & 1 & 3 \\ 1 & 2 & 2 \\ 3 & 1 & 0 \end{vmatrix} \overset{\{R_1 \leftrightarrow R_2\}}{=} - \begin{vmatrix} 1 & 2 & 2 \\ 0 & 1 & 3 \\ 3 & 1 & 0 \end{vmatrix} \overset{\{R_3 - 3R_1\}}{=}$

$$-\begin{vmatrix} 1 & 2 & 2 \\ 0 & 1 & 3 \\ 0 & -5 & -6 \end{vmatrix} \overset{\{R_3+5R_2\}}{=} \begin{vmatrix} 1 & 2 & 2 \\ 0 & 1 & 3 \\ 0 & 0 & 9 \end{vmatrix} = -9.$$

5. $\operatorname{adj}(A) = \begin{bmatrix} 4 & -2 \\ -3 & 1 \end{bmatrix}$; $\det(A) = -2$; $A^{-1} = (-1/2)\begin{bmatrix} 4 & -2 \\ -3 & 1 \end{bmatrix}$.

7. $\operatorname{adj}(A) = \begin{bmatrix} 0 & 1 & -1 \\ -2 & 1 & 0 \\ 1 & -1 & 1 \end{bmatrix}$; $\det(A) = 1$; $A^{-1} = \operatorname{adj}(A)$.

9. $\operatorname{adj}(A) = \begin{bmatrix} -4 & 2 & 0 \\ 1 & 0 & -1 \\ 1 & -2 & 1 \end{bmatrix}$; $\det(A) = -2$;

$$A^{-1} = (-1/2)\begin{bmatrix} -4 & 2 & 0 \\ 1 & 0 & -1 \\ 1 & -2 & 1 \end{bmatrix}.$$

11. $W(x) = \begin{vmatrix} 1 & x & x^2 \\ 0 & 1 & 2x \\ 0 & 0 & 2 \end{vmatrix} = 2$. Since $W(0) = 2$ the given set of functions is linearly independent.

13. $W(x) = \begin{vmatrix} 1 & \cos^2 x & \sin^2 x \\ 0 & -2\cos x \sin x & 2\cos x \sin x \\ 0 & 2\sin^2 x - 2\cos^2 x & 2\cos^2 x - 2\sin^2 x \end{vmatrix} =$

$4\cos x \sin x \cos 2x \begin{vmatrix} 1 & \cos^2 x & \sin^2 x \\ 0 & -1 & 1 \\ 0 & -1 & 1 \end{vmatrix} = 0$. The Wronskian gives no information, but $1 - \cos^2 x - \sin^2 x = 0$ so the set is linearly dependent.

15. Note that $x \mid x \mid = x^2$ for $x \geq 0$ and $x \mid x \mid = -x^2$ for $x < 0$. Therefore $W(x) = \begin{vmatrix} x^2 & x^2 \\ 2x & 2x \end{vmatrix} = 0$ if $x \geq 0$ and $W(x) = \begin{vmatrix} x^2 & -x^2 \\ 2x & -2x \end{vmatrix}$ when $x < 0$. Since $W(x) = 0$ for all $x, -1 \leq x \leq 1$, the Wronskian test is inconclusive. Thus suppose that $c_1 x^2 + c_2 x \mid x \mid = 0$ for all $x, -1 \leq x \leq 1$. Then for $x = 1$ we have $c_1 + c_2 = 0$ and for $x = -1$ we obtain $c_1 - c_2 = 0$. It follows that $c_1 = c_2 = 0$ and the set $\{x^2, x \mid x \mid\}$ is linearly independent.

17. The column operations $C_1 \leftrightarrow C_2, C_3 - 3C_1, C_3 + 2C_2$ reduce A to the matrix $L = \begin{bmatrix} 1 & 0 & 0 \\ 2 & 1 & 0 \\ 2 & 2 & -1 \end{bmatrix}$. Therefore $Q = E_1 E_2 E_3$ where $E_1 = \begin{bmatrix} 0 & 1 & 0 \\ 1 & 0 & 0 \\ 0 & 0 & 1 \end{bmatrix}$, $E_2 = \begin{bmatrix} 1 & 0 & -3 \\ 0 & 1 & 0 \\ 0 & 0 & 1 \end{bmatrix}$,

and $E_3 = \begin{bmatrix} 1 & 0 & 0 \\ 0 & 1 & 2 \\ 0 & 0 & 1 \end{bmatrix}$.

Multiplication yields $Q = \begin{bmatrix} 0 & 1 & 2 \\ 1 & 0 & -3 \\ 0 & 0 & 1 \end{bmatrix}$. It is easily seen that

$\det(Q) = \det(Q^{\mathrm{T}}) = -1$.

19. The column operations $C_2 - 2C_1, C_3 + C_1, C_3 + 4C_4$ transform A to $L = \begin{bmatrix} 1 & 0 & 0 \\ 3 & -1 & 0 \\ 4 & -8 & -26 \end{bmatrix}$.

If $E_1 = \begin{bmatrix} 1 & -2 & 0 \\ 0 & 1 & 0 \\ 0 & 0 & 1 \end{bmatrix}$, $E_2 = \begin{bmatrix} 1 & 0 & 1 \\ 0 & 1 & 0 \\ 0 & 0 & 1 \end{bmatrix}$,

$E_3 = \begin{bmatrix} 1 & 0 & 0 \\ 0 & 1 & 4 \\ 0 & 0 & 1 \end{bmatrix}$, and $Q = E_1E_2E_3 = \begin{bmatrix} 1 & -2 & -7 \\ 0 & 1 & 4 \\ 0 & 0 & 1 \end{bmatrix}$

then $AQ = L$ and $\det(Q) = \det(Q^{\mathrm{T}}) = 1$.

21. $\det(A(x)) = x^2 + 1 > 0$ for all real x. $\operatorname{adj}(A(x)) = \begin{bmatrix} x & -1 \\ 1 & x \end{bmatrix}$ so $A^{-1} = [1/(x^2 + 1)]\begin{bmatrix} x & -1 \\ 1 & x \end{bmatrix}$.

23. $\det(A(x)) = 4x^2 + 8 > 0$ for all real x. $\operatorname{adj}(A) =$

$\begin{bmatrix} x^2+4 & -2x & x^2 \\ 2x & 4 & -2x \\ x^2 & 2x & x^2+4 \end{bmatrix}$ so $A^{-1} = [1/(4x^2 + 8)] \operatorname{adj}(A)$.

25. $\det(L) = 1$ so $L^{-1} = \operatorname{adj}(L) = \begin{bmatrix} 1 & 0 & 0 \\ -a & 1 & 0 \\ ac-b & -c & 1 \end{bmatrix}$. $\det(U) = 1$ so $U^{-1} = \operatorname{adj}(U) =$

$\begin{bmatrix} 1 & -a & ac-b \\ 0 & 1 & -c \\ 0 & 0 & 1 \end{bmatrix}$.

Chapter 6

Eigenvalues and Applications

6.1 Quadratic Forms

1. $A = \begin{bmatrix} 2 & 2 \\ 2 & -3 \end{bmatrix}.$

3. $A = \begin{bmatrix} 1 & 1 & -3 \\ 1 & -4 & 4 \\ -3 & 4 & 3 \end{bmatrix}.$

5. $A = \begin{bmatrix} 2 & 2 \\ 2 & 1 \end{bmatrix}.$

7. $q(\mathbf{x}) = \mathbf{x}^T A\mathbf{x}$ where $A = \begin{bmatrix} 2 & 3 \\ 3 & 2 \end{bmatrix}$. A has eigenvalues $\lambda_1 = 5, \lambda_2 = -1$ with corresponding eigenvectors $a\begin{bmatrix} 1 \\ 1 \end{bmatrix}, b\begin{bmatrix} 1 \\ -1 \end{bmatrix}$, respectively, where $a \neq 0$ and $b \neq 0$. In particular $Q = (1/\sqrt{2})\begin{bmatrix} 1 & 1 \\ 1 & -1 \end{bmatrix}$. The form is indefinite.

9. $q(\mathbf{x}) = \mathbf{x}^T A\mathbf{x}$ for $A = \begin{bmatrix} 1 & 2 & 2 \\ 2 & 1 & 2 \\ 2 & 2 & 1 \end{bmatrix}$. The eigenvalues for A are $\lambda_1 = 5$ and $\lambda_2 = -1$ (algebraic multiplicity 2). An eigenvector for $\lambda_1 = 5$ is $\mathbf{u_1} = [1, 1, 1]^T$. The vectors $\mathbf{w_2} = [-1, 1, 0]^T$ and $\mathbf{w_3} = [-1, 0, 1]^T$

   are eigenvectors for $\lambda_2 = -1$. The Gram-Schmidt process yields orthogonal eigenvectors $\mathbf{u_2} = \mathbf{w_2} = [-1, 1, 0]^T$ and $\mathbf{u_3} = [-1, -1, 2]^T$.

We form Q by normalizing the set $\{\mathbf{u_1}, \mathbf{u_2}, \mathbf{u_3}\}$ of eigenvectors;

$$Q = \begin{bmatrix} 1/\sqrt{3} & -1/\sqrt{2} & -1/\sqrt{6} \\ 1/\sqrt{3} & 1/\sqrt{2} & -1/\sqrt{6} \\ 1/\sqrt{3} & 0 & 2/\sqrt{6} \end{bmatrix}.$$

The form is indefinite.

11. $q(\mathbf{x}) = \mathbf{x}^T A\mathbf{x}$ where $A = \begin{bmatrix} 3 & -1 \\ -1 & 3 \end{bmatrix}$. A has eigenvalues $\lambda_1 = 2$ and $\lambda_2 = 4$ with corresponding eigenvectors $\mathbf{u_1} = [1, 1]^T$ and $\mathbf{u_2} =$ $[-1, 1]^T$, respectively. The set $\{\mathbf{u_1}, \mathbf{u_2}\}$ is orthogonal. We normalize $\mathbf{u_1}$ and $\mathbf{u_2}$ to obtain Q; $Q = (1/\sqrt{2})\begin{bmatrix} 1 & -1 \\ 1 & 1 \end{bmatrix}$. The form is positive definite.

13. Set $q(\mathbf{x}) = 2x^2 + \sqrt{3}\,xy + y^2$. Then $q(\mathbf{x}) = \mathbf{x}^T A\mathbf{x}$ for $A =$ $\begin{bmatrix} 2 & \sqrt{3}/2 \\ \sqrt{3}/2 & 1 \end{bmatrix}$. A has eigenvalues $\lambda_1 = 1/2, \lambda_2 = 5/2$ with corresponding eigenvectors $\mathbf{u_1} = [-1, \sqrt{3}]^T$ and $\mathbf{u_2} = [\sqrt{3}, 1]^T$, respect- ively. Since $\{\mathbf{u_1}, \mathbf{u_2}\}$ is an orthogonal set we may normalize and

obtain$Q = \begin{bmatrix} -1/2 & \sqrt{3}/2 \\ \sqrt{3}/2 & 1/2 \end{bmatrix}$. The substitution $\mathbf{x} = Q\mathbf{y}$ yields

$q(\mathbf{x}) = (1/2)u^2 + (5/2)v^2 = 10$. The graph corresponds to the ellipse $u^2/20 + v^2/4 = 1$.

15. Set $q(\mathbf{x}) = x^2 + 6xy - 7y^2$. Then $q(\mathbf{x}) = \mathbf{x}^T A\mathbf{x}$ where $A = \begin{bmatrix} 1 & 3 \\ 3 & -7 \end{bmatrix}$.

A has eigenvalues $\lambda_1 = -8$ and $\lambda_2 = 2$ with corresponding eigenvectors $\mathbf{u_1} = [-1, 3]^T$ and $\mathbf{u_2} = [3, 1]^T$, respectively. Since $\{\mathbf{u_1}, \mathbf{u_2}\}$ is an orthogonal set we may normalize to obtain $Q = (1/\sqrt{10})\begin{bmatrix} -1 & 3 \\ 3 & 1 \end{bmatrix}$.

The substitution $\mathbf{x} = Q\mathbf{y}$ yields $q(\mathbf{x}) = -8u^2 + 2v^2 = 8$. The graph corresponds to the hyperbola $v^2/4 - u^2 = 1$.

17. If $q(\mathbf{x}) = xy$ then $q(\mathbf{x}) = \mathbf{x}^T A\mathbf{y}$ where $A = \begin{bmatrix} 0 & 1 \\ 1 & 0 \end{bmatrix}$. The mat- rix A is diagonalized by $Q = (1/\sqrt{2})\begin{bmatrix} 1 & -1 \\ 1 & 1 \end{bmatrix}$ and $Q^T AQ = D = \begin{bmatrix} 1 & 0 \\ 0 & -1 \end{bmatrix}$. The substitution $\mathbf{x} = Q\mathbf{y}$ yields $q(\mathbf{x}) = u^2 - v^2 = 4$. The graph corresponds to the hyperbola $u^2/4 - v^2/4 = 1$.

19. If $q(\mathbf{x}) = 3x^2 - 2xy + 3y^2$ then $q(\mathbf{x}) = \mathbf{x}^T A\mathbf{x}$ where $A = \begin{bmatrix} 3 & -1 \\ -1 & 3 \end{bmatrix}$.

A is diagonalized by $Q = (1/\sqrt{2})\begin{bmatrix} -1 & 1 \\ 1 & 1 \end{bmatrix}$ and $Q^T AQ = D$ where $D = \begin{bmatrix} 4 & 0 \\ 0 & 2 \end{bmatrix}$. The transformed equation is the ellipse $4u^2 + 2v^2 = 16$, or $u^2/4 + v^2/8 = 1$.

6.2 Systems of Differential Equations

1. The given system has matrix equation $\mathbf{x}'(t) = A\mathbf{x}(t)$ where

$\mathbf{x}(t) = [u(t), v(t)]^T$ and $A = \begin{bmatrix} 5 & -2 \\ 6 & -2 \end{bmatrix}$. The eigenvalues for A

are $\lambda_1 = 1$ and $\lambda_2 = 2$ and the corresponding eigenvectors are $\mathbf{u_1} = [1, 2]^T, \mathbf{u_2} = [2, 3]^T$. Thus $\mathbf{x_1}(t) = e^t\mathbf{u_1}$ and $\mathbf{x_2}(t) = e^{2t}\mathbf{u_2}$ are solutions. The general solution is given by $\mathbf{x}$ $(t) = b_1\mathbf{x_1}(t) +$

$b_2\mathbf{x_2}(t)$; that is $\mathbf{x}(t) = b_1 e^t \begin{bmatrix} 1 \\ 2 \end{bmatrix} + b_2 e^t \begin{bmatrix} 2 \\ 3 \end{bmatrix}$. It is easily seen that

$\mathbf{x_0} = \mathbf{u_1} + 2\mathbf{u_2}$ so the solution $\mathbf{x}(t) = e^t \begin{bmatrix} 1 \\ 2 \end{bmatrix} + 2e^{2t}\begin{bmatrix} 2 \\ 3 \end{bmatrix} = \begin{bmatrix} e^t + 4e^{2t} \\ 2e^t + 6e^{2t} \end{bmatrix}$ satisfies the

initial condition.

3. The system is $\mathbf{x}'(t) = A\mathbf{x}(t)$ where $A = \begin{bmatrix} 1 & 1 \\ 2 & 2 \end{bmatrix}$. A has eigen- values $\lambda_1 = 0$ and $\lambda_2 = 3$ with corresponding eigenvectors $\mathbf{u_1} =$

$[-1, 1]^T$ and $\mathbf{u_2} = [1, 2]^T$. The general solution is given by $\mathbf{x}(t) =$

$b_1 \begin{bmatrix} -1 \\ 1 \end{bmatrix} + b_2 e^{3t}\begin{bmatrix} 1 \\ 2 \end{bmatrix}$. The particular solution that satisfies the

initial condition is $\mathbf{x}(t) = -3\begin{bmatrix} -1 \\ 1 \end{bmatrix} + 2e^{3t}\begin{bmatrix} 1 \\ 2 \end{bmatrix} = \begin{bmatrix} 3 + 2e^{3t} \\ -3 + 4e^{3t} \end{bmatrix}$.

5. The system is $\mathbf{x}'(t) = A\mathbf{x}(t)$ where $A = \begin{bmatrix} 0.5 & 0.5 \\ -0.5 & 0.5 \end{bmatrix}$. A has eigenvalues $\lambda_1 = 0.5 + 0.5i$ and $\lambda_2 = 0.5 - 0.5i$ with corresponding eigenvectors $\mathbf{u_1} = [-i, 1]^T$ and $\mathbf{u_2} = [i, 1]^T$. The general solution is $\mathbf{x}(t) = b_1 e^{\lambda_1 t}\mathbf{u_1} + b_2 e^{\lambda_2 t}\mathbf{u_2}$ where $e^{\lambda_1 t} = e^{(0.5+0.5i)t} =$ $e^{t/2}[\cos(t/2) + i\sin(t/2)]$ and $e^{\lambda_2 t} = e^{(0.5-0.5i)t} = e^{t/2}[\cos(t/2) - i\sin(t/2)]$. The equation $\mathbf{x_0} = b_1\mathbf{u_1} + b_2\mathbf{u_2}$ has solution $b_1 = 2 + 2i$ and $b_2 = 2 - 2i$ so the particular solution that satisfies the initial condition is $\mathbf{x}(t) = 4e^{(t/2)}\begin{bmatrix} \cos(t/2) + \sin(t/2) \\ \cos(t/2) - \sin(t/2) \end{bmatrix}$.

$$2ie^{(4-2i)t}\begin{bmatrix} -2+2i \\ 1 \end{bmatrix} = 4e^{4t}\begin{bmatrix} 2\cos 2t + 4\sin 2t \\ -\sin 2t \end{bmatrix}.$$

7. The system is $\mathbf{x}'(t) = A\mathbf{x}(t)$ where $A = \begin{bmatrix} 4 & 0 & 1 \\ -2 & 1 & 0 \\ -2 & 0 & 1 \end{bmatrix}$. A has eigenvalues $\lambda_1 = 1, \lambda_2 = 2,$ and $\lambda_3 = 3$ with corresponding eigenvectors $\mathbf{u_1} = [0,1,0]^{\mathrm{T}}, \mathbf{u_2} = [1,-2,-2]^{\mathrm{T}},$ and $\mathbf{u_3} = [-1,1,1]^{\mathrm{T}}$, respectively. Therefore the general solution is $\mathbf{x}(t) =$

$b_1e^t\begin{bmatrix} 0 \\ 1 \\ 0 \end{bmatrix} + b_2e^{2t}\begin{bmatrix} 1 \\ -2 \\ -2 \end{bmatrix} + b_3e^{3t}\begin{bmatrix} -1 \\ 1 \\ 1 \end{bmatrix}$. Since $\mathbf{x_0} = \mathbf{u_1} + \mathbf{u_2} + 2\mathbf{u_3}$ the solution $\mathbf{x}$

$(t) = e^t\begin{bmatrix} 0 \\ 1 \\ 0 \end{bmatrix} + e^{2t}\begin{bmatrix} 1 \\ -2 \\ -2 \end{bmatrix} + 2e^{3t}\begin{bmatrix} -1 \\ 1 \\ 1 \end{bmatrix}$ satisfies the initial condition.

9. (a) The system is $\mathbf{x}'(t) = A\mathbf{x}(t)$ for $A = \begin{bmatrix} 1 & -1 \\ 1 & 3 \end{bmatrix}$. A has only one eigenvalue, $\lambda = 2$, with corresponding eigenvector $u = [1,-1]^{\mathrm{T}}$.

   Therefore $\mathbf{x_1}(t) = e^{2t}\begin{bmatrix} 1 \\ -1 \end{bmatrix}$ is a solution for the system.

   (b) Set $\mathbf{x_2}(t) = te^{\lambda t}\mathbf{u} + e^{\lambda t}\mathbf{y_0}$. Then $\mathbf{x_2}'(t) = e^{\lambda t}(t\lambda\mathbf{u} + \mathbf{u} + \lambda\mathbf{y_0})$ whereas $A\mathbf{x_2}(t) = e^{\lambda t}(t\lambda\mathbf{u} + A\mathbf{y_0})$. Therefore we require that $A\mathbf{y_0} = \mathbf{u} + \lambda\mathbf{y_0}$; that is $(A - \lambda I)\mathbf{y_0} = \mathbf{u}$. One choice is $\mathbf{y_0} =$ $[-2,1]^{\mathrm{T}}$. Thus $\mathbf{x_2}(t) = te^{2t}\begin{bmatrix} 1 \\ -1 \end{bmatrix} + e^{2t}\begin{bmatrix} -2 \\ 1 \end{bmatrix}$ is a solution.

   (c) If $\mathbf{y}(t) = c_1\mathbf{x_1}(t) + c_2\mathbf{x_2}(t)$ note that $\mathbf{y}(0) = c_1\mathbf{u} + c_2\mathbf{y_0}$. Since $\{\mathbf{u}, \mathbf{y_0}\}$ is a linearly independent set for every $\mathbf{x_0}$ in R^2 we may find c_1 and c_2 such that $\mathbf{x_0} = c_1\mathbf{u} + c_2\mathbf{y_0}$

6.3 Transformation to Hessenberg Form

1. The desired elementary row operation is $R_3 - 4R_2$. Performing this operation on the (3 x 3) identity matrix yields $Q_1 = \begin{bmatrix} 1 & 0 & 0 \\ 0 & 1 & 0 \\ 0 & -4 & 1 \end{bmatrix}$.

$Q_1^{-1} = \begin{bmatrix} 1 & 0 & 0 \\ 0 & 1 & 0 \\ 0 & 4 & 1 \end{bmatrix}$ and $Q_1AQ_1^{-1} = H = \begin{bmatrix} -7 & 16 & 3 \\ 8 & 9 & 3 \\ 0 & 1 & 1 \end{bmatrix}$.

3. Let Q_1 denote the permutation matrix $Q_1 = \begin{bmatrix} 1 & 0 & 0 \\ 0 & 0 & 1 \\ 0 & 1 & 0 \end{bmatrix}$. Then $Q_1 = Q_1^{-1}$ and Q_1A interchanges the second and third rows of A. Further $(Q_1A)Q_1$ interchanges the second and third columns of Q_1A. Therefore $H = Q_1AQ_1^{-1} = \begin{bmatrix} 1 & 1 & 3 \\ 1 & 3 & 1 \\ 0 & 4 & 2 \end{bmatrix}$.

5. The desired elementary row operation is $R_3 + 3R_2$. Performing this operation on the (3 x 3) identity matrix yields $Q_1 = \begin{bmatrix} 1 & 0 & 0 \\ 0 & 1 & 0 \\ 0 & 3 & 1 \end{bmatrix}$. $Q_1^{-1} = \begin{bmatrix} 1 & 0 & 0 \\ 0 & 1 & 0 \\ 0 & -3 & 1 \end{bmatrix}$ and $H = Q_1AQ_1^{-1} = \begin{bmatrix} 3 & 2 & -1 \\ 4 & 5 & -2 \\ 0 & 20 & -6 \end{bmatrix}$.

7. Performing the elementary row operations $R_3 - R_2$ and $R_4 - R_2$ on the (4 x 4) identity matrix yields $Q_1 = \begin{bmatrix} 1 & 0 & 0 & 0 \\ 0 & 1 & 0 & 0 \\ 0 & -1 & 1 & 0 \\ 0 & -1 & 0 & 1 \end{bmatrix}$. $Q_1^{-1} = \begin{bmatrix} 1 & 0 & 0 & 0 \\ 0 & 1 & 0 & 0 \\ 0 & 1 & 1 & 0 \\ 0 & 1 & 0 & 1 \end{bmatrix}$ and $H = Q_1AQ_1^{-1} = \begin{bmatrix} 1 & -3 & -1 & -1 \\ -1 & -1 & -1 & -1 \\ 0 & 0 & 2 & 0 \\ 0 & 0 & 0 & 2 \end{bmatrix}$.

9. If $Q_1 = \begin{bmatrix} 1 & 0 & 0 & 0 \\ 0 & 0 & 0 & 1 \\ 0 & 0 & 1 & 0 \\ 0 & 1 & 0 & 0 \end{bmatrix}$ then $Q_1 = Q_1^{-1}$ and Q_1A interchanges the second and fourth rows of A whereas $(Q_1A)Q_1$ interchanges the second and fourth columns Of Q_1A. Therefore

$$Q_1AQ_1 = \begin{bmatrix} 1 & 3 & 1 & 2 \\ 1 & 2 & 0 & 2 \\ 0 & 1 & 1 & 3 \\ 0 & 2 & 1 & 1 \end{bmatrix}.$$

Now the desired elementary row operation is $R_4 - 2R_3$ so set $Q_2 = \begin{bmatrix} 1 & 0 & 0 & 0 \\ 0 & 1 & 0 & 0 \\ 0 & 0 & 1 & 0 \\ 0 & 0 & -2 & 1 \end{bmatrix}$.

Then $Q_2^{-1} = \begin{bmatrix} 1 & 0 & 0 & 0 \\ 0 & 1 & 0 & 0 \\ 0 & 0 & 1 & 0 \\ 0 & 0 & 2 & 1 \end{bmatrix}$ and

$$H = Q_2Q_1AQ_1^{-1}Q_2^{-1} = \begin{bmatrix} 1 & 3 & 5 & 2 \\ 1 & 2 & 4 & 2 \\ 0 & 1 & 7 & 3 \\ 0 & 0 & -11 & -5 \end{bmatrix}.$$

13. $p(t) = (t-2)^3(t+2)$ is the characteristic polynomial for H. Since H and A are similar, $p(t)$ is also the characteristic polynomial for A. Therefore A has eigenvalues $\lambda_1 = 2$ and $\lambda_2 = -2$ and $\lambda_1 = 2$ has algebraic (and hence geometric) multiplicity 3 (cf. Exercise 12).

15. $[\mathbf{e_1}, \mathbf{e_2}, \mathbf{e_3}], [\mathbf{e_1}, \mathbf{e_3}, \mathbf{e_2}], [\mathbf{e_2}, \mathbf{e_1}, \mathbf{e_3}], [\mathbf{e_2}, \mathbf{e_3}, \mathbf{e_1}], [\mathbf{e_3}, \mathbf{e_1}, \mathbf{e_2}],$
$[\mathbf{e_3}, \mathbf{e_2}, \mathbf{e_1}]$.

17. There are $n!$ $(n \times n)$ permutation matrices.

6.4 Eigenvalues of Hessenberg Matrices

1. Note that the given matrix H is in unreduced Hessenberg form. We have $\mathbf{w_0} = \mathbf{e_1} = [1,0]^T, \mathbf{w_1} = H\mathbf{w_0} = [2,1]^T$, and $\mathbf{w_2} = H\mathbf{w_1} = [4,3]^T$. The vector equation $a_0\mathbf{w_0} + a_1\mathbf{w_1} = -\mathbf{w_2}$ is equivalent to the system

$$\begin{aligned} a_0 + 2a_1 &= -4 \\ a_1 &= -3 \end{aligned}.$$

The system has solution $a_0 = 2, a_1 = -3$ so $p(t) = 2 - 3t + t^2$.

3. Note that the given matrix H is in unreduced Hessenberg form. We have $\mathbf{w_0} = \mathbf{e_1} = [1,0,0]^T, \mathbf{w_1} = H\mathbf{w_0} = [1,2,0]^T, \mathbf{w_2} = H\mathbf{w_1} = [1,4,2]^T$, and $\mathbf{w_3} = H\mathbf{w_2} = [3,6,8]^T$. The vector equation
$a_0\mathbf{w_0} + a_1\mathbf{w_1} + a_2\mathbf{w_2} = -\mathbf{w_3}$ is equivalent to the system of equations

$$\begin{aligned} a_0 + a_1 + a_2 &= -3 \\ 2a_1 + 4a_2 &= -6 \\ 2a_2 &= -8 \end{aligned}.$$

The system has unique solution $a_0 = -4, a_1 = 5, a_2 = -4$ so $p(t) = -4 + 5t - 4t^2 + t^3$.

5. Note that the given matrix H is in unreduced Hessenberg form. We have $\mathbf{w_0} = \mathbf{e_1} = [1,0,0]^T, \mathbf{w_1} = H\mathbf{w_0} = [2,1,0]^T, \mathbf{w_2} = H\mathbf{w_1} = [8,3,1]^T$, and $\mathbf{w_3} = H\mathbf{w_2} = [29,14,8]^T$. The vector equation $a_0\mathbf{w_0} + a_1\mathbf{w_1} + a_2\mathbf{w_2} = -\mathbf{w_3}$ is equivalent to the system of equations

$$\begin{array}{rcrcrcr} a_0 & + & 2a_1 & + & 8a_2 & = & -29 \\ & & a_1 & + & 3a_2 & = & -14 \\ & & & & a_2 & = & -8 \end{array} .$$

The system has unique solution $a_0 = 15, a_1 = 10, a_2 = -8$ so $p(t) = 15 + 10t - 8t^2 + t^3$.

7. Note that the given matrix H is in unreduced Hessenberg form. We have $\mathbf{w_0} = \mathbf{e_1} = [1,0,0,0]^T, \mathbf{w_1} = H\mathbf{w_0} = [0,1,0,0]^T$, $\mathbf{w_2} = H\mathbf{w_1} = [1,2,1,0]^T, \mathbf{w_3} = H\mathbf{w_2} = [2,6,2,2]^T$, and $\mathbf{w_4} = H\mathbf{w_3} = [8,18,8,6]^T$. The vector equation $a_0\mathbf{w_0} + a_1\mathbf{w_1} + a_2\mathbf{w_2} + a_3\mathbf{w_3} = -\mathbf{w_4}$ is equivalent to the system of equations

$$\begin{array}{rcrcrcrcr} a_1 & & & + & a_2 & + & 2a_3 & = & -8 \\ & & a_1 & + & 2a_2 & + & 6a_3 & = & -18 \\ & & & & a_2 & + & 2a_3 & = & -8 \\ & & & & & & 2a_3 & = & -6 \end{array}$$

The system has unique solution $a_0 = 0, a_1 = 4, a_2 = -2, a_3 = -3$, so $p(t) = 4t - 2t^2 - 3t^3 + t^4$.

9. $H = \begin{bmatrix} B_{11} & B_{12} \\ \mathcal{O} & B_{22} \end{bmatrix}$ where $B_{11} = \begin{bmatrix} 1 & -1 \\ 1 & 3 \end{bmatrix}, B_{12} = \begin{bmatrix} 1 & 4 \\ -2 & 1 \end{bmatrix}$ and $B_{22} = \begin{bmatrix} 2 & -1 \\ -1 & 2 \end{bmatrix}$. B_{11} has eigenvalue $\lambda_1 = 2$ (with algebraic multiplicity 2) with corresponding eigenvector $\mathbf{u_1} = [-1,1]^T$.B_{22} has eigenvalues $\lambda_2 = 1, \lambda_3 = 3$ with corresponding eigenvectors $\mathbf{v_2} = [1,1]^T$ and $\mathbf{v_3} = [-1,1]^T$, respectively. Thus H has eigenvalues $\lambda_1 = 2, \lambda_2 = 1, \lambda_3 = 3$. The vector $\mathbf{x_1} = \begin{bmatrix} \mathbf{u_1} \\ \theta \end{bmatrix} = [-1,1,0,0]^T$ is an eigenvector for H corresponding to $\lambda_1 = 2$. The system of equations $(B_{11} - I)\mathbf{u} = -B_{12}\mathbf{v_2}$ has solution $\mathbf{u_2} = [-9,5]^T$, so $\mathbf{x_2} = \begin{bmatrix} \mathbf{u_2} \\ \mathbf{v_2} \end{bmatrix} = [-9,5,1,1]^T$ is an eigenvector of H corresponding to $\lambda_2 = 1$. Similarly $(B_{11} - 3I)\mathbf{u} = -B_{12}\mathbf{v_3}$ has solution $\mathbf{u_3} = [-3,9]^T$ so $\mathbf{x_3} = \begin{bmatrix} \mathbf{u_3} \\ \mathbf{v_3} \end{bmatrix} = [-3,9,-1,1]^T$ is an eigenvector of H corresponding to $\lambda_3 = 3$.

11. $H = \begin{bmatrix} B_{11} & B_{12} \\ \mathcal{O} & B_{22} \end{bmatrix}$ where $B_{11} = \begin{bmatrix} -2 & 0 & -2 \\ -1 & 1 & -2 \\ 0 & 1 & -1 \end{bmatrix}, B_{12} = \begin{bmatrix} 1 \\ 3 \\ -2 \end{bmatrix}$, and $B_{22} = [2]$. B_{11} has eigenvalues $\lambda_1 = 0$ and $\lambda_2 = -1$ (algebraic multiplicity 2) with corresponding eigenvectors $\mathbf{u_1} = [-1,1,1]^T$ and $\mathbf{u_2} = [-2,0,1]^T$.$B_{22}$ has eigenvalue $\lambda_3 = 2$

with corresponding eigenvector $\mathbf{v_3} = [1]$. Thus H has eigenvalues $\lambda_1 = 0, \lambda_2 = -1$, and $\lambda_3 = 2$. The vectors $\mathbf{x_1} = \begin{bmatrix} \mathbf{u_1} \\ \theta \end{bmatrix} = [-1,1,1,0]^{\mathrm{T}}$ and $\mathbf{x_2} = \begin{bmatrix} \mathbf{u_2} \\ \theta \end{bmatrix} = [-2,0,1,0]^{\mathrm{T}}$ are eigenvectors for H corresponding to $\lambda_1 = 0$ and $\lambda_2 = -1$, respectively. The system of equations $(B_{11} - 2I)\mathbf{u_1} = -B_{12}\mathbf{v_3}$ has solution $\mathbf{u_3} = [1/6, 15/6, 1/6]^{\mathrm{T}}$ so $\mathbf{x_3} = \begin{bmatrix} \mathbf{u_3} \\ \mathbf{v_3} \end{bmatrix} =$ $[1/6, 15/6, 1/6, 1]^{\mathrm{T}}$ is an eigenvector for H corresponding to $\lambda_3 = 2$.

15. $P = [\mathbf{e_2}, \mathbf{e_3}, \mathbf{e_1}]$.

6.5 Householder Transformations

1. $Q\mathbf{x} = \mathbf{x} - \gamma\mathbf{u}$ where $\gamma = 2\mathbf{u}^{\mathrm{T}}\mathbf{x}/\mathbf{u}^{\mathrm{T}}\mathbf{u} = (-2)(2)/4 = -1$. Thus $Q\mathbf{x} = [4,1,6,7]^{\mathrm{T}}$.

3. Set $\gamma_1 = 2\mathbf{u}^{\mathrm{T}}\mathbf{A_1}/\mathbf{u}^{\mathrm{T}}\mathbf{u} = -1$ and $\gamma_2 = 2\mathbf{u}^{\mathrm{T}}\mathbf{A_2}/\mathbf{u}^{\mathrm{T}}\mathbf{u} = -2$. Then $Q\mathbf{A_1} = \mathbf{A_1} - \gamma_1\mathbf{u}$ $= [3,5,5,1]^{\mathrm{T}}$ and $Q\mathbf{A_2} = \mathbf{A_2} - \gamma_2\mathbf{u} = [3,1,4,2]^{\mathrm{T}}$.

   Therefore $QA = \begin{bmatrix} 3 & 3 \\ 5 & 1 \\ 5 & 4 \\ 1 & 2 \end{bmatrix}$.

5. Set $\gamma = 2\mathbf{u}^{\mathrm{T}}\mathbf{x}/\mathbf{u}^{\mathrm{T}}\mathbf{u} = -1$. Then $Q\mathbf{x} = \mathbf{x} - \gamma\mathbf{u} = \mathbf{x} + \mathbf{u} = [4,1,3,4]^{\mathrm{T}}$.
   Thus $\mathbf{x}^{\mathrm{T}}Q = (Q\mathbf{x})^{\mathrm{T}} = [4,1,3,4]$.

7. Set $\mathbf{x} = [2,1,2,1]^{\mathrm{T}}$ and $\mathbf{y} = [1,0,1,4]^{\mathrm{T}}$. Then $QA^{\mathrm{T}} = Q[\mathbf{x},\mathbf{y}] =$ $[Q\mathbf{x}, Q\mathbf{y}] = \begin{bmatrix} 1 & 2 \\ 2 & -1 \\ 1 & 2 \\ 2 & 3 \end{bmatrix}$. Therefore $AQ = (QA^{\mathrm{T}})^{\mathrm{T}} =$ $\begin{bmatrix} 1 & 2 & 1 & 2 \\ 2 & -1 & 2 & 3 \end{bmatrix}$.

9. Set $u_1 = 0$. If $a = -\sqrt{4+4+1} = -3$ then $u_2 = v_2 - a = 2 + 3 = 5$. Finally take $u_3 = v_3 = 2$ and $u_4 = v_4 = 1$. Thus $\mathbf{u} = [0,5,2,1]^{\mathrm{T}}$.

11. $a = -\sqrt{4^2 + 3^2} = -5; u_1 = u_2 = 0; u_3 = v_3 - a = 4 + 5 = 9; u_4 = v_4 = 3$. Therefore $\mathbf{u}$ $= [0,0,9,3]^{\mathrm{T}}$.

13. $a = \sqrt{(-3)^2 + 4^2} = 5; u_1 = u_2 = u_3 = 0; u_4 = v_4 - a = -8; u_5 = v_5 = 4$. Therefore $\mathbf{u}$ $= [0,0,0,-8,4]^{\mathrm{T}}$.

15. We want $QA_1 = [1, a, 0]^T$. Therefore $a = -\sqrt{3^2+4^2} = -5$ and $\mathbf{u} = [u_1, u_2, u_3]^T$ where $u_1 = 0, u_2 = 3-(-5) = 8,$ and $u_3 = 4$. Then $\mathbf{u} = [0, 8, 4]^T$.

17. We want $Q\mathbf{A_1} = \begin{bmatrix} 0 \\ a \\ 0 \end{bmatrix}$. Therefore $a = \sqrt{(-4)^2+3^2} = 5$ and $\mathbf{u} = [u_1, u_2, u_3]^T$ where $u_1 = 0, u_2 = -4-5 = -9,$ and $u_3 = 3$. Thus $\mathbf{u} = [0, -9, 3]^T$.

19. We want $Q\mathbf{A_2} = \begin{bmatrix} 1 \\ 4 \\ a \\ 0 \end{bmatrix}$ so $a = \sqrt{(-3)^2+4^2} = 5$. $\mathbf{u} = [u_1, u_2, u_3, u_4]^T$ where $u_1 = u_2 = 0, u_3 = -3-5 = -8,$ and $u_4 = 4$. Thus $\mathbf{u} = [0, 0, -8, 4]^T$.

6.6 QR Factorization & Least-Squares

1. $\mathbf{x}^*$ is the unique solution to $R\mathbf{x} = \mathbf{c}$, where $R = \begin{bmatrix} 1 & 2 \\ 0 & 1 \end{bmatrix}$ and $\mathbf{c} = \begin{bmatrix} 3 \\ 1 \end{bmatrix}$. Thus $\mathbf{x}^* = [1, 1]^T$.

3. $\mathbf{x}^*$ is the unique solution to $R\mathbf{x} = \mathbf{c}$ where $R = \begin{bmatrix} 1 & 2 & 1 \\ 0 & 1 & 3 \\ 0 & 0 & 2 \end{bmatrix}$ and $\mathbf{c} = \begin{bmatrix} 6 \\ 7 \\ 4 \end{bmatrix}$. Thus $\mathbf{x}^* = [2, 1, 2]^T$.

5. We require that $S\mathbf{A_1} = [a, 0]^T$. Therefore $a = \pm\sqrt{a_{11}^2 + a_{21}^2} = -5, u_1 = a_{11} - a = 8,$ and $u_2 = a_{21} = 4$. Consequently $\mathbf{u} = [8, 4]^T$ and $S\mathbf{A_1} = \mathbf{A_1} - \mathbf{u}$ $= [-5, 0]^T$. $S\mathbf{A_2} = \mathbf{A_2} - 2\mathbf{u} = [-11, 2]^T$, so $SA = R = \begin{bmatrix} -5 & -11 \\ 0 & 2 \end{bmatrix}$.

7. We require that $S\mathbf{A_1} = \begin{bmatrix} a \\ 0 \end{bmatrix}$ so take $a = -\sqrt{a_{11}^2 + a_{21}^2} = -4$,

$u_1 = a_{11} - a = 4$, and $u_2 = a_{21} = 4$. Thus $\mathbf{u} = [4, 4]^T$ and $S\mathbf{A_1} =$

$\mathbf{A_1} - \mathbf{u} = [-4, 0]^T$. Also $S\mathbf{A_2} = \mathbf{A_2} - 2\mathbf{u} = [-6, -2]^T$. Therefore $SA = R = \begin{bmatrix} -4 & -6 \\ 0 & -2 \end{bmatrix}$.

9. We require that $S\mathbf{A_2} = [2, a, 0]^T$ so set $a = -\sqrt{a_{22}^2 + a_{32}^2} = -1; u_1 = 0, u_2 = a_{22} - a = 1$, and $u_3 = a_{32} = 1$. Therefore $\mathbf{u} = [0, 1, 1]^T, S\mathbf{A_1} = \mathbf{A_1}, S\mathbf{A_2} = \mathbf{A_2} - \mathbf{u} = [2, -1, 0]^T$, and $S\mathbf{A_3} = \mathbf{A_3} - 14\mathbf{u} = [1, -8, -6]^T$. Consequently $SA = R = \begin{bmatrix} 1 & 2 & 1 \\ 0 & -1 & -8 \\ 0 & 0 & -6 \end{bmatrix}$.

11. We first require Q_1 such that $Q_1\mathbf{A_1} = [a, 0, 0, 0]^T$ so take $\mathbf{u_1} = [6, 2, 2, 4]^T$. Then $Q_1\mathbf{A_1} = \mathbf{A_1} - \mathbf{u_1} = [-5, 0, 0, 0]^T$ and $Q_1\mathbf{A_2} =$
$\mathbf{A_2} - 3\mathbf{u} = [-59/3, 6, 2, 3]^T$. We now require Q_2 such that
$Q_2(Q_1\mathbf{A_2}) = [-59/3, a, 0, 0]^T$. Thus set $\mathbf{u_2} = [0, 13, 2, 3]^T$. Then
$Q_2(Q_1\mathbf{A_1}) = Q_1\mathbf{A_1}$ and $Q_2(Q_1\mathbf{A_2}) = Q_1\mathbf{A_2} - \mathbf{u_2} =$
$[-59/3, -7, 0, 0]^T$. Therefore $Q_2Q_1A = \begin{bmatrix} -5 & -59/3 \\ 0 & -7 \\ 0 & 0 \\ 0 & 0 \end{bmatrix}$.

13. We require a matrix Q such that $Q\mathbf{A_2} = [4, a, 0, 0]^T$. Thus $\mathbf{u} = [0, 8, 0, 4]^T$ and $QA = \begin{bmatrix} 2 & 4 \\ 0 & -5 \\ 0 & 0 \\ 0 & 0 \end{bmatrix}$.

15. Let $Q_1, \mathbf{u_1}, Q_2, \mathbf{u_2}$ be as in Exercise 11. Then $Q_1\mathbf{b} = \mathbf{b} - \mathbf{u_1} = [-5, 8, -2, -3]^T$ and $Q_2(Q_1\mathbf{b}) = Q_1\mathbf{b} - \mathbf{u_2} = [-5, -5, -4, -6]^T$.

The least-squares solution is the unique solution $\mathbf{x}^*$ to $R\mathbf{x} = \mathbf{c}$ where $R = \begin{bmatrix} -5 & -59/3 \\ 0 & -7 \end{bmatrix}$
and $\mathbf{c} = [-5, -5]^T$. Thus $\mathbf{x}^* =$
$[-38/21, 15/21]^T$.

17. With Q and $\mathbf{u}$ as in Exercise 13, $Q\mathbf{b} = \mathbf{b} - (12/5)\mathbf{u} =$
$[2, -56/5, 16, -8/5]^T$. Therefore $\mathbf{x}^*$ is the unique solution to
$R\mathbf{x} = \mathbf{c}$ where $R = \begin{bmatrix} 2 & 4 \\ 0 & -5 \end{bmatrix}$ and $\mathbf{c} = [2, -56/5]^T$. Solving
yields $\mathbf{x}^* = [-87/25, 56/25]^T$.

19. Write $[\mathbf{A_1}, \mathbf{A_2}, \ldots, \mathbf{A_n}]$, where $\{\mathbf{A_1}, \mathbf{A_2}, \ldots, \mathbf{A_n}\}$ is a linearly independent subset of R^n. Now $SA = [S\mathbf{A_1}, S\mathbf{A_2}, \ldots, S\mathbf{A_n}]$. Suppose $c_1, c_2, \ldots, c_n$ are scalars such that $\theta = c_1 S\mathbf{A_1} + c_2 S\mathbf{A_2} + \cdots + c_n S\mathbf{A_n}$.

Then $\theta = S(c_1\mathbf{A_1} + c_2\mathbf{A_2} + \cdots + c_n\mathbf{A_n})$ and S is nonsingular. Therefore $\theta = c_1\mathbf{A_1} + c_2\mathbf{A_2} + \cdots + c_n\mathbf{A_n}$. It follows that $c_1 = c_2 = \cdots = c_n = 0$ and hence, the set $\{S\mathbf{A_1}, S\mathbf{A_2}, \ldots, S\mathbf{A_n}\}$ is linearly independent. For each $j, 1 \le j \le n, S\mathbf{A_j} = \begin{bmatrix} \mathbf{R_j} \\ \theta \end{bmatrix}$ where $\mathbf{R_j}$ is the jth column of R and θ is in R^{m-n}. Therefore the set $\{\mathbf{R_1}, \mathbf{R_2}, \ldots, \mathbf{R_n}\}$ is linearly independent in R^n and the matrix R is nonsingular.

6.7 Matrix Polynomials & The Cayley-Hamilton Theorem

1. $q(A) = A^2 - 4A + 3I = \begin{bmatrix} -1 & 0 \\ 0 & -1 \end{bmatrix}$; $q(B) = B^2 - 4B + 3I = \begin{bmatrix} 0 & 0 \\ 0 & 0 \end{bmatrix}$; $q(C) = C^2 - 4C + 3I = \begin{bmatrix} 15 & -2 & 14 \\ 5 & -2 & 10 \\ -1 & -4 & 6 \end{bmatrix}$.

3. (a) $q(t) = s(t)p(t) + r(t)$ where $s(t) = t^3 + t - 1$ and $r(t) = t + 2$.

   (b) $q(B) = s(B)p(B) + r(B) = r(B)$ since $p(B) = \mathcal{O}$. Thus $q(B) = B + 2I = \begin{bmatrix} 4 & -1 \\ -1 & 4 \end{bmatrix}$.

6.8 Generalized Eigenvectors & Differential Equations

1. (a) The given matrix H has characteristic polynomial $p(t) = (t-2)^2$, so $\lambda = 2$ is the only eigenvalue and it has algebraic multiplicity 2. The vector $\mathbf{v_1} = [1, -1]^{\mathrm{T}}$ is an eigenvector corresponding to $\lambda = 2$. If we solve the system of equations $(H - 2I)\mathbf{x} = \mathbf{v_1}$ we see that $\mathbf{x} = [-1 - a, a]^{\mathrm{T}}$, where a is arbitrary. Taking $a = 0$ we obtain a generalized eigenvector $\mathbf{v_2} = [-1, 0]^{\mathrm{T}}$.

   (b) The given matrix H has characteristic polynomial $p(t) = t(t+1)^2$. The eigenvalue $\lambda = -1$ has corresponding eigenvector $\mathbf{v_1} = [-2, 0, 1]^{\mathrm{T}}$. Solving the system $(H - (-1)I)\mathbf{x} = \mathbf{v_1}$ yields $\mathbf{x} = [2 - 2a, 1, a]^{\mathrm{T}}$ where a is arbitrary. Thus $\mathbf{v_2} = [0, 1, 1]^{\mathrm{T}}$ is a generalized eigenvector for $\lambda = -1$. The eigenvalue $\lambda = 0$ has corresponding eigenvector $\mathbf{w_1} = [-1, 1, 1]^{\mathrm{T}}$.

   (c) The given matrix H has characteristic polynomial $p(t) = (t-1)^2(t+1)$. The eigenvalue $\lambda = 1$ has corresponding eigenvector $\mathbf{v_1} = [-2, 0, 1]^{\mathrm{T}}$. Solving $(H - I)\mathbf{x} = \mathbf{v_1}$ yields $\mathbf{x} = [(5/2) - 2a, 1/2, a]^{\mathrm{T}}$, where a is arbitrary. Thus

$\mathbf{v_2} = [5/2, 1/2, 0]^{\mathrm{T}}$ is a generalized eigenvector of $\lambda = 1$. The eigenvalue $\lambda = -1$ has corresponding eigenvector $\mathbf{w_1} = [-9, -1, 1]^{\mathrm{T}}$.

3. (a) If $Q = \begin{bmatrix} 1 & 0 & 0 \\ 0 & 1 & 0 \\ 0 & 3 & 1 \end{bmatrix}$ then $Q^{-1} = \begin{bmatrix} 1 & 0 & 0 \\ 0 & 1 & 0 \\ 0 & -3 & 1 \end{bmatrix}$ and $H =$ $QAQ^{-1} = \begin{bmatrix} 8 & -69 & 21 \\ 1 & -10 & 3 \\ 0 & -4 & 1 \end{bmatrix}$ is in unreduced Hessenberg form. H has characteristic polynomial $p(t) = (t+1)^2(t-1)$. The eigenvalue $\lambda = -1$ has corresponding eigenvector $\mathbf{v_1} = [3, 1, 2]^{\mathrm{T}}$.

Solving the system $(H - (-1)I)\mathbf{u} = \mathbf{v_1}$ yields $\mathbf{u} =$ $[-(7/2) + (3/2)a, (-1/2) + (1/2)a, a]^{\mathrm{T}}$, where a is arbitrary.
Therefore $\mathbf{v_2} = [-2, 0, 1]^{\mathrm{T}}$ is a generalized eigenvector for $\lambda = -1$. The eigenvalue $\lambda = 1$ has corresponding eigenvector $\mathbf{w_1} = [-3, 0, 1]^{\mathrm{T}}$. Set $\mathbf{y}(t) = Q\mathbf{x}(t)$ and $\mathbf{y_0} = Q\mathbf{x_0} = [-1, -1, -2]^{\mathrm{T}}$.

The system $\mathbf{y}' = H\mathbf{y}$ has general solution $\mathbf{y}(t) = c_1 e^{-t}\mathbf{v_1} +$ $c_2 e^{-t}(\mathbf{v_2} + t\mathbf{v_1}) + c_3 e^{t}\mathbf{w_1}$ and $\mathbf{y_0} = \mathbf{y}(0) = c_1\mathbf{v_1} + c_2\mathbf{v_2} + c_3\mathbf{w_1}$. Solving we obtain $c_1 = -1, c_2 = 2, c_3 = -2$, so $\mathbf{y}(t) =$
$\begin{bmatrix} e^{-t}(6t-7) + 6e^t \\ e^{-t}(2t-1) \\ e^{-t}(4t) - 2e^t \end{bmatrix}$. Therefore

$$\mathbf{x}(t) = Q^{-1}\mathbf{y}(t) = \begin{bmatrix} e^{-t}(6t-7) + 6e^t \\ e^{-t}(2t-1) \\ e^{-t}(-2t+3) - 2e^t \end{bmatrix}.$$

(b) If $Q = \begin{bmatrix} 1 & 0 & 0 \\ 0 & 1 & 0 \\ 0 & 3 & 1 \end{bmatrix}$ then $Q^{-1} = \begin{bmatrix} 1 & 0 & 0 \\ 0 & 1 & 0 \\ 0 & -3 & 1 \end{bmatrix}$ and
$H = QAQ^{-1} = \begin{bmatrix} 2 & 4 & -1 \\ -3 & -4 & 1 \\ 0 & 3 & -1 \end{bmatrix}$ is in unreduced Hessenberg form. H has characteristic polynomial $p(t) = (t+1)^3$. The eigenvalue $\lambda = -1$ has corresponding eigenvector $\mathbf{v_1} = [1, 0, 3]^{\mathrm{T}}$. The system $(H - (-1)I)\mathbf{u} = \mathbf{v_1}$ has solution $\mathbf{u} = [-1 + (1/3)a, 1, a]^{\mathrm{T}}$, where a is arbitrary. Therefore $\mathbf{v_2} = [0, 1, 3]^{\mathrm{T}}$ is a generalized eigenvector of order 2 for $\lambda = -1$. The system $(H - (-1)I)\mathbf{u} = \mathbf{v_2}$ has solution $\mathbf{u} = [(-4/3) + (1/3)a, 1, a]^{\mathrm{T}}$ so $\mathbf{v_3} = [0, 1, 4]^{\mathrm{T}}$ is a generalized eigenvector of order 3 for $\lambda = -1$.
Set $\mathbf{y}(t) = Q\mathbf{x}(t)$ and $\mathbf{y_0} = Q\mathbf{x_0} = [-1, -1, -2]^{\mathrm{T}}$. The system $\mathbf{y}' = H\mathbf{y}$ has general solution $\mathbf{y}(t) = c_1 e^{-t}\mathbf{v_1} + c_2 e^{-t}(\mathbf{v_1} + t\mathbf{v_1}) + c_3 e^{-t}(\mathbf{v_3} + t\mathbf{v_2} + (t^2/2)\mathbf{v_1})$ and

$\mathbf{y_0} = \mathbf{y}(0) = c_1\mathbf{v_1} + c_2\mathbf{v_2} + c_3\mathbf{v_3}$. Solving we obtain $c_1 = -1, c_2 = -5, c_3 = 4$, so $\mathbf{y}(t) = \begin{bmatrix} e^{-t}(2t^2 - 5t - 1) \\ e^{-t}(4t - 1) \\ e^{-t}(6t^2 - 3t - 2) \end{bmatrix}$. Therefore $\mathbf{x}(t) = Q^{-1}\mathbf{y}(t) = \begin{bmatrix} e^{-t}(2t^2 - 5t - 1) \\ e^{-t}(4t - 1) \\ e^{-t}(6t^2 - 15t + 1) \end{bmatrix}$.

(c) If $Q = \begin{bmatrix} 1 & 0 & 0 \\ 0 & 1 & 0 \\ 0 & 3 & 1 \end{bmatrix}$ then $Q^{-1} = \begin{bmatrix} 1 & 0 & 0 \\ 0 & 1 & 0 \\ 0 & -3 & 1 \end{bmatrix}$ and $H =$

$QAQ^{-1} = \begin{bmatrix} 1 & 4 & -1 \\ -3 & -5 & 1 \\ 0 & 3 & -2 \end{bmatrix}$ is in unreduced Hessenberg form. H has characteristic polynomial $p(t) = (t+2)^3$. The eigenvalue $\lambda = -2$ has eigenvector $\mathbf{v_1} = [1, 0, 3]^{\mathrm{T}}$, generalized eigenvector $\mathbf{v_2} = [0, 1, 3]^{\mathrm{T}}$ of order two, and generalized eigenvector $\mathbf{v_3} = [0, 1, 4]^{\mathrm{T}}$ of order 3.

Set $\mathbf{y}(t) = Q\mathbf{x}(t)$ and $\mathbf{y_0} = Q\mathbf{x_0} = [-1, -1, -2]^{\mathrm{T}}$. The system $\mathbf{y}' = H\mathbf{y}$ has general solution $\mathbf{y}(\mathbf{t}) = e^{-2t}[c_1\mathbf{v_1} + c_2(\mathbf{v_2} + t\mathbf{v_1}) + c_3(\mathbf{v_3} + t\mathbf{v_2} + (t^2/2)\mathbf{v_1})]$ and $\mathbf{y_0} = \mathbf{y}(\mathbf{0}) = c_1\mathbf{v_1} + c_2\mathbf{v_2} + c_3\mathbf{v_3}$. Solving yields $c_1 = -1, c_2 = -5$, and $c_3 = 4$, so

$$\mathbf{y}(t) = \begin{bmatrix} e^{-2t}(2t^2 - 5t - 1) \\ e^{-2t}(4t - 1) \\ e^{-2t}(6t^2 - 3t - 2) \end{bmatrix}.$$ Therefore

$$\mathbf{x}(t) = Q^{-1}\mathbf{y}(t) = \begin{bmatrix} e^{-2t}(2t^2 - 5t - 1) \\ e^{-2t}(4t - 1) \\ e^{-2t}(6t^2 - 15t + 1) \end{bmatrix}.$$

5. We see that from part(c) of Exercise 1 that $\mathbf{x}(t) =$
$c_1e^t\mathbf{v_1} + c_2e^t(\mathbf{v_2} + t\mathbf{v_1}) + c_3e^{-t}\mathbf{w_1} =$

$$c_1e^t\begin{bmatrix} -2 \\ 0 \\ 1 \end{bmatrix} + c_2e^t\left(\begin{bmatrix} 5/2 \\ 1/2 \\ 0 \end{bmatrix} + t\begin{bmatrix} -2 \\ 0 \\ 1 \end{bmatrix}\right) + c_3e^{-t}\begin{bmatrix} -9 \\ -1 \\ 1 \end{bmatrix}.$$

$$\begin{matrix} 0 & 2 & 1 & -2 \\ 3 & 5 & -5 & 1 \\ 2 & 4 & -2 & 2 \end{matrix} \xrightarrow{R_3 \leftrightarrow R_1} \begin{matrix} 2 & 4 & -2 & 2 \\ 3 & 5 & -5 & 1 \\ 0 & 2 & 1 & -2 \end{matrix} \xrightarrow{(\frac{1}{2})R_1} \begin{matrix} 1 & 2 & -1 & 2 \\ 3 & 5 & -5 & 1 \\ 0 & 2 & 1 & -2 \end{matrix} \xrightarrow{R_2 - 3R_1} \begin{matrix} 1 & 2 & -1 & 2 \\ 0 & -1 & -2 & -5 \\ 0 & 2 & 1 & -2 \end{matrix}$$

$$\xrightarrow{R_1 + R_2} \begin{matrix} 1 & 4 & 0 & 0 \\ 0 & -1 & -2 & -5 \\ 0 & 2 & 1 & -2 \end{matrix} \xrightarrow{R_3 + 2R_2} \begin{matrix} 1 & 4 & 0 & 0 \\ 0 & 1 & 2 & 5 \\ 0 & 0 & -3 & -12 \end{matrix} \rightarrow \begin{matrix} 1 & 4 & 0 & 0 \\ 0 & 1 & 2 & 5 \\ 0 & 0 & 1 & 4 \end{matrix}$$

$x_1 = 12$ $x_3 = 4$

$x_2 = -3$